New Theory and Design of Ellipsoidal Heads for Pressure Vessels

Jinyang Zheng · Keming Li

New Theory and Design of Ellipsoidal Heads for Pressure Vessels

ZHEJIANG UNIVERSITY PRESS
浙江大学出版社

Springer

Jinyang Zheng
Zhejiang University
Hangzhou, Zhejiang, China

Keming Li
Zhejiang University
Hangzhou, Zhejiang, China

ISBN 978-981-16-0466-9 ISBN 978-981-16-0467-6 (eBook)
https://doi.org/10.1007/978-981-16-0467-6

Jointly published with Zhejiang University Press, China
The print edition is not for sale in China Mainland. Customers from China Mainland please order the print book from: Zhejiang University Press

This Springer imprint is published by the registered company Springer Nature Singapore Pte Ltd.
The registered company address is: 152 Beach Road, #21-01/04 Gateway East, Singapore 189721, Singapore

Preface

Ellipsoidal heads are widely used as the end closures of pressure vessels in the chemical, petroleum, nuclear, marine, aerospace and food processing industries due to their good stress distribution and ease of fabrication. Recently, ellipsoidal heads have become larger and thinner; for example, super-large, super-thin ellipsoidal heads with a diameter of up to 43 m and a diameter-to-thickness ratio of about 1000 are used in the cylindrical steel containment vessels of CAP1400 nuclear power plants. This book is intended to provide comprehensive coverage of stress, failure, design and fabrication of ellipsoidal heads.

Many researchers have conducted extensive research on the analysis and design of ellipsoidal heads under internal pressure, and contributed to the development of design methods in the codes and standards. The initial research was mainly based on elastic stress theory. In 1925, Huggenberger gave the formulas for calculating the membrane stresses of ellipsoidal heads. Coates performed theoretical analysis for bending stresses at the head-cylinder junction. Using Coates's theoretical method, Maker found that the ratio of maximum stress in the head to circumferential stress in the cylinder shows a function of the radius-to-height ratio of heads, as stated in the book *Process Equipment Design: Vessel Design* by Brownell and Young. Based on Maker's study, a proposed curve on the function was used as the stress-intensification factor by the American Society of Mechanical Engineers (ASME) Boiler and Pressure Vessel Code (BPVC), Section VIII. This curve was then displaced with a classic formula which was used for preventing plastic collapse of ellipsoidal heads in the current pressure vessel codes and standards such as the ASME BPVC, Section VIII, Division 1 (ASME VIII-1) and the Chinese standard *Pressure Vessels—Part 3: Design* (GB/T 150.3). It should be noted that elastic theory is still used to prevent the plastic collapse of head crowns with an equivalent spherical shell formula in the current codes and standards such as the ASME BPVC, Section VIII, Division 2 (ASME VIII-2) and the European Standard *Unfired Pressure Vessels—Part 3: Design* (EN 13445-3).

Subsequently, perfectly plastic theory was used to prevent the plastic collapse of the ellipsoidal heads. In 1961, Shield and Drucker used the upper and lower bound theorems of limit analysis to develop a formula for calculating the limit pressure of torispherical heads. From the 1970s to 2007, design curves based on the Shield and Drucker formula were used in ASME VIII-2 to prevent the plastic collapse of torispherical heads and 2:1 ellipsoidal heads. These design curves are also used in the Chinese industry standard *Steel Pressure Vessels—Design by Analysis* (JB 4732). Kalnins and Updike used a perfectly plastic material model and large deformation theory to develop design curves for preventing the plastic collapse of torispherical heads, and a design method based on the work of Kalnins and Updike with some modification has been used to prevent the knuckle yielding of torispherical and ellipsoidal heads in EN 13445-3 since 2002.

In addition, ellipsoidal heads may fail by buckling, especially large low-pressure heads, due to the existence of compressive circumferential stresses at the head knuckle even if the heads are subjected to internal pressure. The buckling of ellipsoidal and torispherical heads was intensively investigated by Bushnell, Galletly, Miller, et al. Bushnell developed the famous BOSOR5 program for the buckling of elastic-plastic shells of revolution, and made important contributions to the computerized buckling analysis of shells. Galletly developed a design formula to prevent buckling of torispherical heads on the basis of perfectly plastic theory, which was subsequently used to prevent the buckling of ellipsoidal heads in EN 13445-3. Miller developed formulas to prevent the buckling of torispherical heads on the basis of elastic theory. Miller's formulas were used to prevent the buckling of ellipsoidal heads in ASME VIII-1 since 2001. Since 2007, ASME VIII-2 has used Miller's formulas in the design of ellipsoidal heads. It should be noted that most of the researchers have focused on the buckling of torispherical heads, and ellipsoidal heads are designed as equivalent torispherical heads in the current codes and standards. Ellipsoidal heads have apparently not been studied as extensively as torispherical heads. In this book, we will conduct deep research on the buckling of ellipsoidal heads.

In 2007, the elastic-plastic analysis method was introduced to ASME VIII-2 and used in the design-by-analysis of pressure components. The effects of material strain hardening and geometric nonlinearity are considered in the elastic-plastic analysis method. However, the effects of both the material strain hardening and geometric nonlinearity are not included in current formulas for the design of ellipsoidal heads. In fact, the load-carrying capacity of ellipsoidal heads will be strengthened by deforming toward hemispherical ones. The ultimate load-carrying capacity of ellipsoidal heads is not fully developed using the current design formulas based on the elastic or perfectly-plastic theory. In this book, the effects of material strain hardening and geometric nonlinearity will be considered in the analysis and design of ellipsoidal heads.

In Chaps. 2 and 3, the buckling and plastic collapse behavior of ellipsoidal heads, respectively, are comprehensively investigated using both the nonlinear finite element method and experiments. In particular, modern measurement technologies such as 3D laser scanning are used to obtain the buckling behavior of ellipsoidal heads with a diameter of up to 5000 mm. It is found that the buckling of ellipsoidal heads under internal pressure has three characteristics: locality, progressivity and self-limitation. In addition, material strain hardening has a significant effect on the plastic collapse pressure of ellipsoidal heads. New formulas are developed, respectively, for predicting the buckling and plastic collapse pressure of ellipsoidal heads, and are found to be in good agreement with the experimental data. This lays the foundation for the development of design rules for ellipsoidal heads. Subsequently, in Chap. 4, we develop a new failure mechanism-based method for the design of ellipsoidal heads based on elastic-plastic theory. This new design method can fully develop the pressure-carrying capacity of ellipsoidal heads compared to the methods in the current codes and standards, which reduces the head wall thickness and thus the cost.

Generally, fabrication quality affects the load-carrying capacity or in-service performance of ellipsoidal heads; for example, shape deviation arising during head fabrication always reduces the buckling pressure of heads. Forming strain also occurs during head fabrication and plastic strain tends to cause phase transformation in materials, significantly affecting the material properties. In particular, metastable austenitic stainless steel can undergo a strain-induced martensitic transformation during the cold forming of heads, which may cause the degradation of its mechanical properties. Therefore, the effects of fabrication quality on capacity or performance need to be considered in the design of ellipsoidal heads. In the end, Chap. 5 covers the development of methods for controlling the fabrication quality of ellipsoidal heads, including a method for evaluating the shape deviation of ellipsoidal heads on the basis of non-contact measurement, a formula for predicting the maximum forming strain and a method for determining the warm forming temperature to avoid strain-induced martensitic transformation. These methods can play a key role in controlling the fabrication quality of ellipsoidal heads.

This book can help to provide a better understanding of the buckling and plastic collapse behavior of ellipsoidal heads under internal pressure. The development of the new design method for ellipsoidal heads based on elastic-plastic theory is an advance over the current rules based on elastic or perfectly-plastic theory, which can lead to a more uniform safety margin and further help to achieve the iso-strength design of cylindrical pressure vessels with ellipsoidal heads. Additionally, controlling the fabrication quality is an essential aspect in the design of the pressure components such as ellipsoidal heads in this book. It can remind designers of fabrication quality effects. This book can be used as a technical reference for researchers and engineers in the fields of engineering mechanics, engineering design, manufacturing engineering and industrial engineering.

We are particularly indebted to Academician Xuedong Chen of the Chinese Academy of Engineering and China National Machinery Industry Corporation for encouragement and advice in the preparation of this book. We gratefully acknowledge the researchers and professors who have contributed to the success of this book: Zekun Zhang, Chaohua Gu, Guoyou Sun, Qunjie Lu, Xiaoping Zhang, Xiao Zhang, Xiaobo Zhu, Yingzhe Wu, Wenzhu Peng and Yehong Yu of Zhejiang University. We also express our gratitude to the engineers who have contributed to the experimental work of this book: Yijun Zhang and Jun Cui of Hefei General Machinery Research Institute Co., Ltd.; Caisheng Gu and Binfeng Zhang of Yixing Hokkai Head Plate Co., Ltd.; Shenghua Liu and Honghui Ge of Shanghai Nuclear

Engineering Research and Design Institute Co., Ltd.; Guanghua Lin, Ming Zhang, Jian Chen, Weixing Huang and Lei Shi of Changzhou Saifu Chemical Engineering Equipment Installation Co., Ltd.; Zhiping Lu and Xiangyang Lü of Changzhou Kuangda Weide Machinery Co., Ltd.; Guofu Liu of Dingjin General Machinery Co., Ltd.; and Weican Guo of Zhejiang Academy of Special Equipment Science.

Hangzhou, China

Jinyang Zheng
Keming Li

Contents

Nomenclature

a	Major radius of an ellipse
A	Percentage elongation after fracture
b	Minor radius of an ellipse
B	Sensitivity of martensitic transformation to forming temperature
C	Coefficient used in the fitting of a new formula for plastic collapse pressure
C_1, C_2	Coefficients used in the torispherical head design
$C_{\alpha'}$	Actual mass fraction of the martensite phase
d	Distance from the center of the circular plate to its radius
D_b	Diameter of the flat plate or the diameter of the intermediate product
D_f	Final outside diameter of a component after forming
D_i	Inside diameter of a head
$D_i/2h_i$	Radius-to-height ratio of ellipsoidal heads, which equals the inside diameter of the head divided by twice the inside height of the head
D_m	Mean diameter of a head
D_n	Nominal inside diameter of a head
D_o	Outside diameter of a head
$D_o/2h_o$	Ratio of the major to the minor axis of ellipsoidal heads, which equals the outside diameter of the head divided by twice the outside height of the head
D_R	Distance from the head center line to the nozzle center line
E	Young's modulus
E_r	Ratio of Young's modulus at room temperature and design temperature
e_{min}	Minimum eigenvalue in elastic buckling analysis
F_D	Design factor
FN	Ferrite number
FN_{sat}	Saturation value of the amount of transformed martensite
G	Constant used in the torispherical head design

h_b	Height of a buckle
h_c	Length of a cylinder
h_i	Inside height of an ellipsoidal head
h_o	Outside height of an ellipsoidal head
h_s	Height of the straight section of a head
h_t	Inside height of a torispherical head
L	Inside crown radius of a torispherical head
LE	Lateral expansion
L_m	Mean crown radius, $L_m = L + 0.5t$
M	Knuckle shakedown factor
M_r	Knuckle to crown stress intensity ratio
K	Stress ratio factor
K_c	Coefficient in Roche and Alix formula for plastic buckling pressure
K_e	Coefficient in Roche and Alix formula for elastic buckling pressure
K_o	Spherical radius factor for external pressurized ellipsoidal heads
KV_2	Absorbed energy
p	Uniformly distributed internal pressure
P	Allowable pressure
P_b	Buckling pressure
P_c	Predicted plastic collapse pressure
p_c	Calculation pressure
P_{cb}	Experimental plastic collapse pressure
P_d	Design pressure
P_e	Internal pressure expected to produce elastic buckling of the knuckle in a torispherical head
P_{exp}	Experimental buckling pressure
P_f	Internal pressure expected to result in a buckling failure of the knuckle in a torispherical head
P_y	Internal pressure expected to result in maximum stress equal to the material yield strength in a torispherical head
P_L	Limit pressure
P_k	Allowable internal pressure of a torispherical head based on a buckling failure of the knuckle
r	Inside knuckle radius of a torispherical head
R_1, R_2	Principal radii of head curvature
R_e	Radius used in the torispherical head design
R_{eff}	Effective pressure radius
R_{eL}	Lower yield strength
R_f	Final mean radius after forming
R_m	Mean radius of a spherical shell
R_{sm}	Smallest mean radius of the segment (mean crown radius, mean knuckle radius of an ellipsoidal head)

R_n	Nozzle inside radius
R_{nc}	Radius of the nozzle opening in the vessel along the long chord, for radial nozzles $R_{nc} = R_n$
R_o	Equivalent outside spherical radius
$R_{p0.2}$	0.2% proof strength
$R_{p1.0}$	1.0% proof strength
R_{om}	Original mean radius, equal to infinity for a flat plate
S	Allowable stress
S_e	Elastic buckling stress
S_a	Alternating stress
S_f	Flow stress
SFE	Stacking fault energy
S_u	Ultimate tensile strength
S_y	Yield strength
t	Thickness of a shell or head
T	Forming temperature in Celsius
t_b	Minimum required thickness to prevent buckling of ellipsoidal heads
t_c	Minimum required thickness to prevent the plastic collapse of ellipsoidal heads
t_m	Average measured thickness of tested heads
t_y	Minimum required thickness to prevent axisymmetric yielding of ellipsoidal heads
x	Head radial distance to meridian
X	Longitudinal coordinate on the outer surface of a head
X_0	Longitudinal coordinate on the blank before stamping
X_{mean}	Mean of the measured longitudinal coordinate on four radiuses
X_{sim}	Simulated longitudinal coordinate
Y_{punch}	Convex die position
Z	Reduction of area of a specimen after fracture
α	Coefficient used in the formula for predicting plastic collapse pressure
α_c	Factor used in the design formula for preventing the plastic collapse of ellipsoidal heads
β	Reduction factor used in the formula for predicting buckling pressure
β_c	Factor used in the design formula for preventing the buckling of ellipsoidal heads
β_t	Factor used in the design equation of torispherical heads in EN 13445-3
β_{th}, ϕ_{th}	Angles used in the torispherical head design
δ_n	Nominal thickness of a head
δ_p	Bulging height
$\delta(\varphi, \theta)$	Normal deviation from perfect shape

ε_f	Calculated forming strain
ε_t	Normal strain in thickness direction
ε_X	Normal strain in longitudinal direction
ε_θ	Normal strain in latitudinal direction
$\bar{\varepsilon}^{pl}_{max}$	Maximum equivalent plastic strain
ν	Poisson's ratio
σ_{eq}	Equivalent stress
σ_φ	Meridional stress
σ_θ	Circumferential stress
ϕ	Weld joint factor
ρ	Curvature radius
φ, φ_1, φ_2, φ_3	Circumferential angle for bulging of weld
θ, θ_1, θ_2, θ_3	Meridional angle for bulging of weld
θ_n	Angle between the nozzle center line and the vessel center line
$\omega(\cdot)$	Mass fraction of chemical composition; for example, ω(Ni) stands for the mass fraction of Ni
Δd	Maximum deviation between ellipsoidal and torispherical heads
ΔFN	Ferrite number transformation
Δh	Height difference between ellipsoidal and torispherical heads
Δ_i	Maximum inside shape deviation
Δ_o	Maximum outside shape deviation
ΔX	Measured longitudinal elongation
ΔX_{sim}	Simulated longitudinal elongation

Chapter 1
Introduction

1.1 General

1.1.1 Head Types

End closures on pressure vessels are termed "heads". The reason for the use of heads on cylindrical vessels arose from the development of power steam boilers in the early nineteenth century [1]. The development of the thermal cracking process in the petroleum industry between 1915 and 1930 resulted in the construction of thousands of pressure vessels with heads [1]. Figure 1.1 shows variant of heads widely used in engineering.

With cylindrical vessel closures, flat or conical heads seem to be a choice because there is no difficulty of dishing. However, discontinuity of shape curvature at the junction between these heads and cylinders causes high edge stress at or near the junction. With the development of dishing process, dished heads (including shallow spherical, hemispherical, torispherical and ellipsoidal heads) are more applicable for the closures of pressure vessels. The major advantage of dished heads (except shallow spherical heads) over flat or conical heads is its large reduction in the discontinuity of shape curvature at the junction between the heads and cylinders, resulting in a reduction of edge stress at or near the junction. Additionally, these dished heads are considerably stronger than flat or conical heads.

Although hemispherical heads are stronger than either torispherical or ellipsoidal heads, they are not widely used because of the excessive forming required during fabrication, resulting in a costly and difficult fabrication process. Majority of cylindrical vessels have used either torispherical or ellipsoidal heads. A torispherical head is composed of a spherical crown, a toroidal knuckle and a very short cylindrical shell, and the three components have common tangent where they meet. The discontinuation of the junction at the spherical crown and toroidal knuckle also leads to edge stresses at or near the junction. An ellipsoidal head is a dished closure made on a truly ellipsoidal former. Edge stress is non-existence due to gradually changing curvature of ellipsoidal shape; thus, an ellipsoidal head is the ideal head choice for

J. Zheng and K. Li, *New Theory and Design of Ellipsoidal Heads for Pressure Vessels*,
https://doi.org/10.1007/978-981-16-0467-6_1

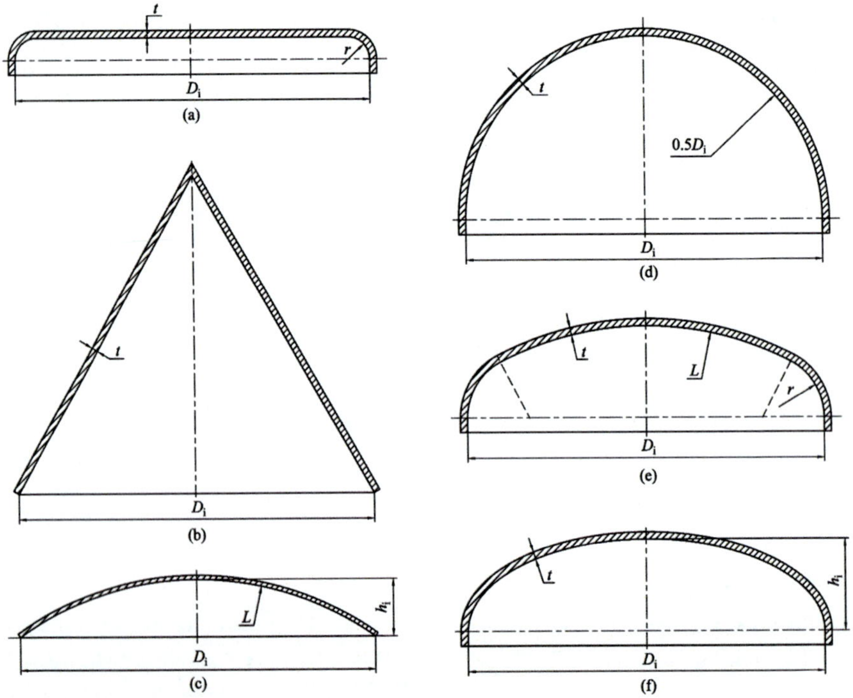

Fig. 1.1 Head shape variants: **a** flat, **b** conical, **c** shallow spherical, **d** hemispherical, **e** torispherical, **f** ellipsoidal

pressure vessels. Subsequently, this book is intended for ellipsoidal heads. Analysis and design of other head types can be seen in the book by Jawad and Farr [2].

1.1.2 Steel and Its Mechanical Properties

Materials used for construction of heads in engineering application include steel, aluminum, copper, titanium and zirconium. The most commonly used material is steel, including carbon as well as low- and high-alloy steel. As such, this book is specifically intended for steel-made ellipsoidal heads.

The typical engineering stress–strain curves for a ductile steel commonly used in engineering constructions are shown in Fig. 1.2a for carbon and low alloy steel and in Fig. 1.2b for high alloy steel. The stress–strain curves are obtained from tensile test of specimens, and the tensile test method is included in standards [3, 4]. The stress–strain curves include elastic range where deformation is relative small and plastic range where strain becomes large. Mechanical properties of steel are described as follows. Elasticity is the property of a material to return to its original

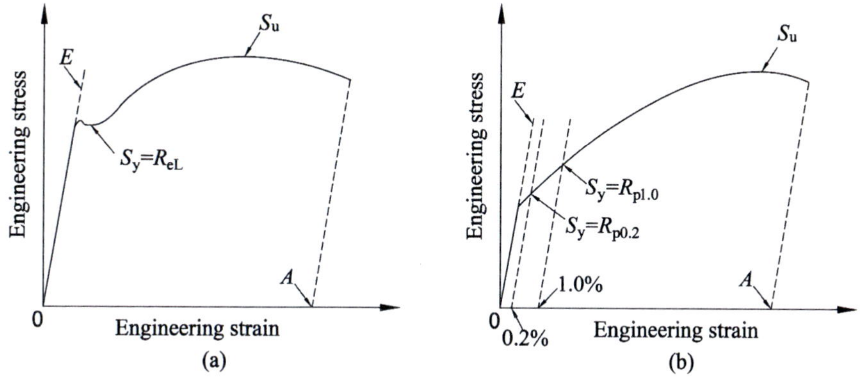

Fig. 1.2 Engineering stress–strain curves: **a** carbon and low alloy steel; **b** high alloy steel

shape after removal of load. The modulus of elasticity, or Young's modulus (E) is the slope of the elastic portion of the stress–strain curves. The value of E for the above ductile steels is about 200 GPa. Yield strength demonstrates the condition where the steel starts yielding. Determination of yield strength depends on its steel types: While carbon and low alloy steel both exhibit yield plateau, yield strength $\left(S_y\right)$ is defined as lower yield strength (R_{eL}), as shown in Fig. 1.2a; For high alloy steel, yield strength $\left(S_y\right)$ is determined using the "Offset Method": 0.2% proof strength $\left(R_{p0.2}\right)$ corresponds to the offset of 0.2% (i.e., permanent deformation equal to 0.2%), and 1.0% proof strength $\left(R_{p1.0}\right)$ corresponds to permanent deformation equal to 1.0%, as shown in Fig. 1.2b. This portion of the curve after yielding is then followed by the well-known phenomenon of strain hardening characterized by progressive increase in stress with large plastic deformation until the stress reaches a maximum value known as ultimate tensile strength (S_u). As stress decreases, the specimen fractures. The elongation after fracture (A) is the increase in the gauge length measured after fracture, expressed as a percentage of the original gauge length. The ductility of a material is characterized by the elongation after fracture as well as the percentage reduction in area. In addition, toughness is the ability of a material to absorb energy during plastic deformation. The detailed mechanical properties of steels can be seen in the book by Harvey [5].

Compare two types of Chinese steel (Q345R—a low alloy steel and S30408—a high alloy steel) which, as an example, are commonly used to construct ellipsoidal heads that have the following mechanical properties. For Q345R plate with a thickness of 16–36 mm, the specified lower yield strength (R_{eL}) is 325 MPa, the specified ultimate tensile strength (S_u) range is 500–630 MPa, and the specified percentage elongation after fracture (A) is more than or equal to 21%, according to the Chinese standard GB/T 713 [6]. For S30408 plate, the specified 0.2 and 1.0% proof strengths ($R_{p0.2}$ and $R_{p1.0}$) are more than or equal to 220 MPa and 250 MPa respectively, the specified minimum ultimate tensile strength is 520 MPa, and the specified percentage

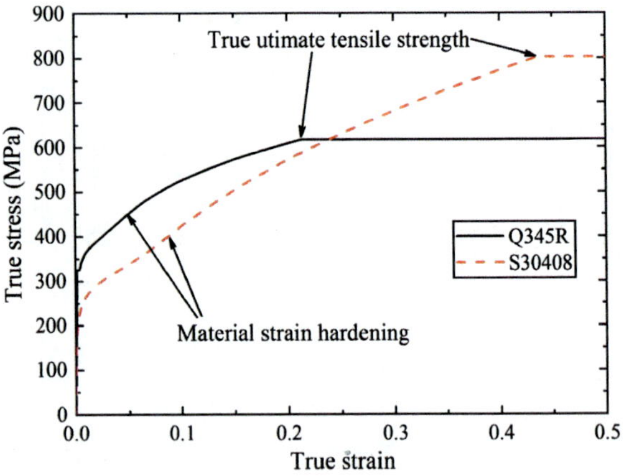

Fig. 1.3 True stress-true strain curves of Q345R and S30408 determined by material model in ASME VIII-2

elongation after fracture is more than or equal to 40%, according to the Chinese standard GB/T 24511 [7].

Generally, linear elastic material and elastic-perfectly plastic material models are used in the analysis and design of pressure vessels and their components. Since 2007, ASME VIII-2 [8] has provided a material model for true stress-true strain curves used in design-by-analysis of pressure vessels when the material strain hardening characteristics are considered; see Formulas (3-D.1)–(3-D.13) in ASME VIII-2. Young's modulus, yield strength and ultimate tensile strength are used to determine a true stress–strain curve, according to the material model in ASME VIII-2. Figure 1.3 shows the true stress-true strain curves for Q345R ($E = 200$ GPa, $S_y = 325$ MPa, $S_u = 500$ MPa) and for S30408 ($E = 195$ GPa, $S_y = 220$ MPa, $S_u = 520$ MPa). The true stress–strain curve beyond a value of true ultimate tensile strength is perfectly plastic.

1.1.3 Application

Ellipsoidal heads can be welded to cylinders or be attached to other components with a bolting flange. Ellipsoidal heads are usually used as closures of internally pressurized vessels in the industries of chemical, petroleum, nuclear, marine, aerospace and food processing. For example, ellipsoidal heads with a diameter of around 1–2 m are usually used in oil storage tanks. Stainless steel ellipsoidal heads with a diameter of about 5 m are used in beer fermenters. A super-large super-thin ellipsoidal head with a diameter of about 40 m and thickness of about 40 mm is used as an end closure of steel containments in nuclear power plants such as AP1000 [9–11] and CAP1400 [12].

1.2 Stress of Ellipsoidal Heads

1.2.1 Elastic Analysis

Membrane theory is used to perform elastic analysis on ellipsoidal heads under internal pressure. Based on membrane theory, meridional stress (σ_φ) and circumferential stress (σ_θ) of shell of revolution are determined using the following formulas:

$$\sigma_\varphi = \frac{pR_2}{2t} \tag{1.1}$$

$$\sigma_\theta = \sigma_\varphi\left(2 - \frac{R_2}{R_1}\right) \tag{1.2}$$

For ellipsoidal shells, R_1 and R_2 are the principal radii of head curvature (Fig. 1.4) and determined by the following formulas:

$$R_1 = \frac{\left[\left(b^2 - a^2\right)x^2 + a^4\right]^{3/2}}{a^4 b} \tag{1.3}$$

$$R_2 = \frac{\left[\left(b^2 - a^2\right)x^2 + a^4\right]^{1/2}}{b} \tag{1.4}$$

Therefore, we can obtain Formulas (1.5) and (1.6) for calculating the meridional stress (σ_φ) and circumferential stress (σ_θ) of ellipsoidal heads under internal pressure:

$$\sigma_\varphi = \frac{pR_2}{2t} = \frac{p}{2t}\frac{\left[\left(b^2 - a^2\right)x^2 + a^4\right]^{1/2}}{b} \tag{1.5}$$

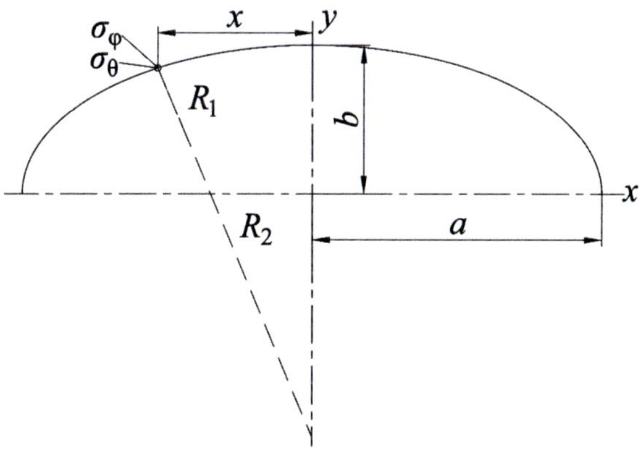

Fig. 1.4 Geometry of ellipsoidal head

$$\sigma_\theta = \frac{p}{2t}\frac{\left[(b^2 - a^2)x^2 + a^4\right]^{1/2}}{b}\left[2 - \frac{a^4}{(b^2 - a^2)x^2 + a^4}\right] \quad (1.6)$$

In 1925, the above formulas were first developed using Huggenberger's stress analysis [13]. Figure 1.5 shows the stress distribution of ellipsoidal heads with $a/b =$ 1.4, 2.0 and 2.5 based on Formulas (1.5) and (1.6) from Huggenberger's analysis. The tensile meridional stress (σ_φ) decreases along head radial distance to meridian (x); while the tensile circumferential stress (σ_θ) decreases along the radial distance (x), and becomes compressive circumferential stress as a/b increases. For head apex where $R_1 = R_2 = a^2/b$, the meridional stress (σ_φ) is equal to circumferential stress (σ_θ), and they are determined by Formula (1.7); the stresses of the head apex are both tensile and increase with the increase in a/b. For head equator where $R_1 = a^2/b$ and $R_2 = a$, σ_φ and σ_θ are determined by Formulas (1.8) and (1.9). According to Formula (1.9), the circumferential stress in head equator $\sigma_\theta < 0$ when $a/b > 1.4$, which means that head knuckle is in compressive state. A detailed stress analysis of ellipsoidal shells can be also seen in the book by Harvey [5].

$$\sigma_\theta = \sigma_\varphi = p\frac{a^2}{2bt} \quad (1.7)$$

$$\sigma_\varphi = p\frac{a}{2t} \quad (1.8)$$

$$\sigma_\theta = p\frac{a}{t}\left(1 - \frac{a^2}{2b^2}\right) \quad (1.9)$$

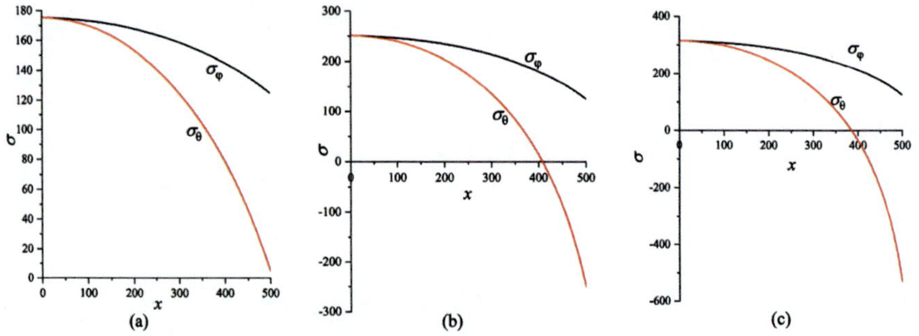

Fig. 1.5 Stress distribution of ellipsoidal heads based on elastic analysis: **a** $a/b = 1.4$; **b** $a/b = 2$; **c** $a/b = 2.5$

1.2.2 Elastic–Plastic Analysis

Regarding standard (2:1) ellipsoidal heads with a radius-to-height ratio $D_i/2h_i$ of 2, it is assumed that a standard ellipsoidal head has a diameter of 1000 mm and thickness of 5 mm. A cylinder, attached to the ellipsoidal head, has the same diameter and thickness as the ellipsoidal head. The ellipsoidal head and cylinder are assumed to be made of the steel Q345R including plastic behavior, and its true stress-true strain curve is shown in Fig. 1.3. We used finite element (FE) method to perform elastic–plastic analysis on the ellipsoidal head.

Figure 1.6 shows the distribution of circumferential and meridional stresses in the middle surfaces of the ellipsoidal head and cylinder under different pressures based on elastic–plastic FE analysis. It is seen that since elastic deformation occurs at low pressure of 2 MPa, the circumferential and meridional stresses gradually changing along the radial distance are very close to Huggenberger's elastic stress analysis results, except for the circumferential stress difference at or near the junction between the head and cylinder due to edge stresses occurring at or near the junction. For this reason, Huggenberger's analysis does not consider the edge effect in the head near the junction with the cylinder. Coates [14], Kraus et al. [15, 16] and Magnucki et al. [17–19] investigated the problem of this edge effect at the junction between ellipsoidal heads and cylinders. Fluctuations of the stresses occur due to local plastic deformation occurring at head crown and knuckle when pressure increases to 4 MPa. As pressure further increases, the head goes into overall plastic state, and stress fluctuations at the head become small, with the exception of edge stress at or near the junction. In addition, with an increase in pressure, tensile circumferential stresses at head crown and meridional stresses increase, while compressive circumferential stresses at head knuckle first increase and then decrease, even become tensile stresses.

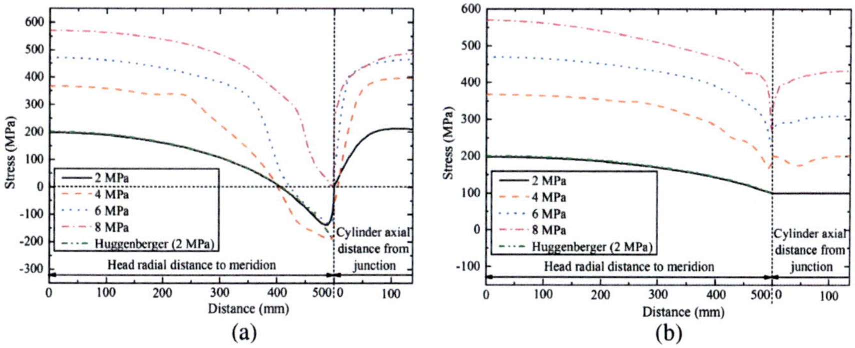

Fig. 1.6 **a** Circumferential and **b** meridional stress distribution of ellipsoidal heads with cylinders based on elastic–plastic analysis

1.2.3 Stress Comparison Between Ellipsoidal and Torispherical Heads

Due to the difficulty of ellipsoidal die in its fabrication, equivalent torispherical heads are used as ellipsoidal heads. According to the four-center method, crown and knuckle radii of equivalent torispherical heads (approximate ellipsoidal heads) are determined by the Formulas (1.10) and (1.11), respectively. The height of equivalent torispherical head is determined by Formula (1.12). Therefore, the geometry of an equivalent torispherical head is obtained by the diameter (D_i) and radius-to-height ratio ($D_i/2h_i$) of an ellipsoidal head.

$$r/D_i = \frac{(D_i/2h_i)^2 + 1 - (D_i/2h_i - 1)\sqrt{(D_i/2h_i)^2 + 1}}{4(D_i/2h_i)^2} \tag{1.10}$$

$$L/D_i = \frac{(D_i/2h_i)^2 + 1 + (D_i/2h_i - 1)\sqrt{(D_i/2h_i)^2 + 1}}{4(D_i/2h_i)} \tag{1.11}$$

$$h_t = L - \sqrt{(L - D_i/2)(L + D_i/2 - 2r)} \tag{1.12}$$

Figure 1.7 shows shape deviation between ellipsoidal and equivalent torispherical heads where radius-to-height ratio is 2, and it is shown that the maximum deviation (Δd) located in head knuckle and the torispherical head deviates outside the ellipsoidal head. Table 1.1 lists maximum deviation (Δd) in head knuckle and height

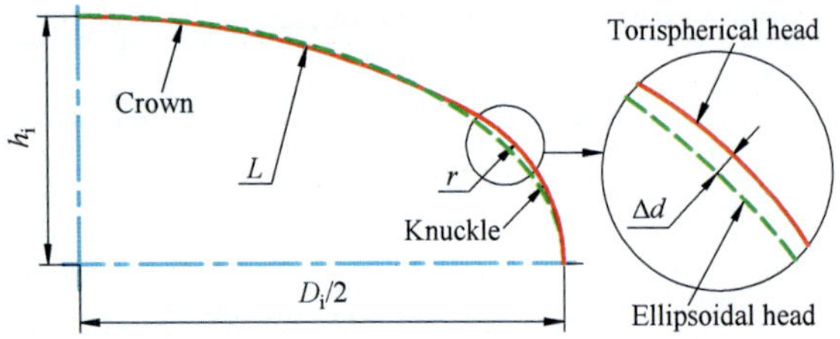

Fig. 1.7 Shape deviation between ellipsoidal and its equivalent torispherical heads (from Ref. [21], with permission of ASME)

Table 1.1 Geometric difference between ellipsoidal and its equivalent torispherical heads

$D_i/2h_i$	Δh(mm)	$\Delta h/h_i$(%)	Δd(mm)	$\Delta d/D_i$(%)
1.4	0.5	0.13	3.7	0.37
1.7	2.5	0.85	7.7	0.77
2	−1.2	−0.46	8.1	0.81
2.2	−0.9	−0.39	9.4	0.94
2.6	−1.9	−0.96	10.5	1.05
3.0	−1.5	−0.89	10.7	1.07

difference (Δh) between ellipsoidal and equivalent torispherical heads with a diameter of 1000 mm when the ratio ($D_i/2h_i$) varies from 1.4 to 3.0. It is shown that compared with an ellipsoidal head, the height of an equivalent torispherical head is greater when the ratio ($D_i/2h_i$) is smaller; while torispherical head is lower when the ratio ($D_i/2h_i$) is larger in value, i.e., $D_i/2h_i \geq 2.0$. Additionally, the maximum deviation (Δd) grows in value as the ratio ($D_i/2h_i$) increases. When $D_i/2h_i = 3.0$, the difference of the maximum deviation reaches a maximum value of 1.07%, which satisfies the shape deviation requirement specified in codes and standards that the inner surface of an ellipsoidal head shall not deviate outside of the specified shape by more than 1.25% of an inside diameter; see Sect. 1.5.2.1.

Figure 1.8 shows the distribution of circumferential and meridional stresses at the middle surface of the above standard ellipsoidal head and its equivalent torispherical head. It is shown that at a low pressure of 2 MPa, the stresses in the ellipsoidal head are gradually changing along the radial distance; while stresses in the equivalent torispherical head show large fluctuation near or at the junction between the head crown and knuckle. In addition, the maximum compressive stress in the equivalent torispherical head is higher than that in the ellipsoidal head, as shown in Fig. 1.8a. The reason is that ellipsoidal head curvature is continuous while torispherical head curvature is not continuous at the crown-knuckle junction, as also discussed by Bushnell [20]. With an increase in pressure, such as the pressure of 8 MPa, the stress differences between ellipsoidal and its torispherical head become very small, as shown in Fig. 1.8b.

Section 1.2.3 is adapted from Ref. [21] with permission of ASME.

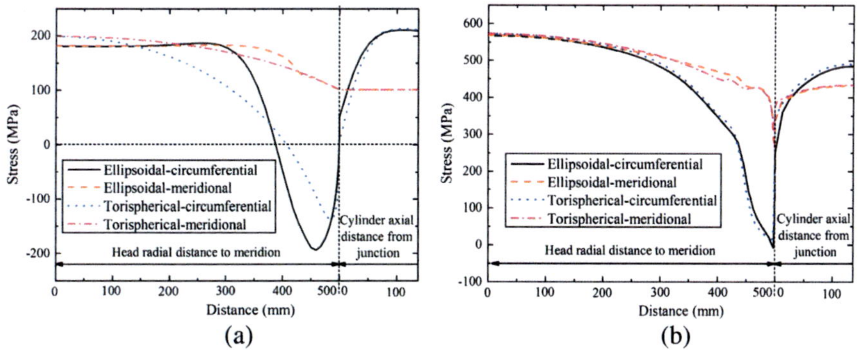

Fig. 1.8 Stress comparison between ellipsoidal and its equivalent torispherical heads under internal pressure: **a** 2 MPa; **b** 8 MPa

1.3 Failure Prediction of Ellipsoidal Heads

1.3.1 Failure Mode

According to stress analysis of ellipsoidal heads in Sect. 1.2, bi-axial tensile stresses in head crown cause the occurrence of plastic collapse, while compressive circumferential stresses in head knuckle cause the occurrence of buckling. Therefore, these two types of failure mode, buckling and plastic collapse, are considered for ellipsoidal heads under static internal pressure [22–24].

Many researchers [22–30] did a lot of research on plastic collapse of ellipsoidal heads, but the effect of material strain hardening was not considered in these research. The initial work was to prevent plastic collapse of heads until some researchers such as Galletly [31], Bushnell [32] pointed out that heads may fail by buckling, especially large low-pressure heads, due to the existence of compressive circumferential stresses at head knuckle even if they are subjected to internal pressure; in particular, the case of buckling of a very large, thin torispherical head occurred under hydrostatic testing reported by Fino and Schneider [33]. Many researchers [20, 22–24, 34–43] did a lot of research on buckling of ellipsoidal heads. In particular, Bushnell [44] developed the famous BOSOR5 program for buckling of elastic–plastic shells of revolution, and made important contribution to the computerized buckling analysis of shells [45]. Blachut et al. [24, 46] presented a detailed review of the study on failure of pressure vessel components including ellipsoidal heads.

Formulas for prediction of plastic collapse and buckling pressures of ellipsoidal heads are summarized as follows.

1.3.2 Plastic Collapse

(1) Formula based on Huggenberger's stress analysis

Based on Huggenberger's stress analysis, the maximum elastic stress occurs in head apex; see Sect. 1.2.1. The maximum elastic stress is determined by Formula (1.7). According to the maximum stress-based failure criterion that head will fail by plastic collapse when the maximum stress (i.e. σ_θ or σ_φ) at head apex reaches yield strength (S_y) of material, we can determine the Formula (1.13) for calculating plastic collapse pressure of ellipsoidal heads by substituting S_y into σ_θ of Formula (1.7).

$$P_c = 4S_y \frac{a}{b} \frac{t}{D_m} \tag{1.13}$$

(2) Formulas based on stress ratio (K)

A study was made by Maker in 1932 with ellipsoidal heads to determine the effect of radius-to-height ratio ($D_i/2h_i$) on the stress level at the head-to-cylinder junction for a constant ratio 32 of cylinder radius to head thickness [1]. The study indicated that the stress ratio (K) of the maximum stress in the head to the circumferential stress in the cylinder changes with radius-to-height ratio. The stress ratio (K) is calculated using Formula (1.14) in codes and standards such as ASME VIII-1 [47] and GB/T 150.3 [48]. According to the maximum stress-based failure criterion, we can obtain the Formulas (1.15) and (1.16) for calculating the collapse pressure of ellipsoidal heads by substituting S_y into allowable stress S of design Formulas (1.27) and (1.31) presented in Sect. 1.4.2.

$$K = \frac{1}{6}\left[2 + (D_i/2h_i)^2\right] \tag{1.14}$$

$$P_c = \frac{2S_y t}{K D_i + 0.2t} \tag{1.15}$$

$$P_c = \frac{2S_y t}{K D_i + 0.5t} \tag{1.16}$$

(3) Galletly formula based on elastic-perfectly plastic theory

Based on numerical analysis using elastic-perfectly plastic material model, Galletly et al. [23] proposed Formula (1.17) for predicting collapse pressure of ellipsoidal heads under internal pressure. The collapse pressure is referred to the axisymmetric yielding pressure determined by twice-elastic-slope method or twice-yield-point-deflection method based on a pressure-apex deformation curve. In addition, this formula is only applicable for $500 < D_i/t < 1500$ and $207\ \text{MPa} < S_y < 414\ \text{MPa}$.

$$P_c = 2.0S_y\frac{t}{D_i}\left(1 + 50\frac{S_y}{E}\right) \tag{1.17}$$

1.3.3 Local Buckling

(1) Tovstik formula

Tovstik [35] derived an analytical Formula (1.18) for elastic buckling pressure of internally pressurized ellipsoidal heads. However, this formula is applicable for heads with a large radius-to-height ratio ($D_i/2h_i \geq 3.3$).

$$P_b = \frac{16E}{\sqrt{3(1-\nu^2)}}\left(\frac{b/a}{a/t}\right)^2\left[1+\sqrt{\frac{193}{32\sqrt{3(1-\nu^2)}}}\sqrt{\frac{1-4(b/a)^2}{b/t}}\right] \tag{1.18}$$

(2) Galletly formulas

Using the BOSOR5 program, Galletly et al. [23, 38] investigated buckling pressure of steel ellipsoidal heads with a radius-to-height ratio of 2, and developed Formulas (1.19) and (1.20) respectively for elastic and plastic buckling pressure of 2:1 ellipsoidal heads subjected to internal pressure. Formula (1.19) is used to predict elastic buckling pressure of 2:1 ellipsoidal heads with a diameter-to-thickness ratio in the range of $750 < D_i/t < 1500$; while Formula (1.20) is used to predict plastic buckling pressure of 2:1 ellipsoidal heads having a diameter-to-thickness ratio in the range of $600 < D_i/t < 1500$.

$$P_b = 455E(D_i/t)^{-2.5} \tag{1.19}$$

$$P_b = 10.4S_y(D_i/t)^{-1.25} \tag{1.20}$$

(3) Roche and Alix formulas

Roche and Alix [39] presented a formula to calculate elastic buckling pressure of an ellipsoidal head under internal pressure. The formula derived from computational results for elastic buckling is

$$P_b = K_e\frac{\pi^2 E}{1-\nu^2}\frac{t^2h_i^2}{(D_i/2)^4} \tag{1.21}$$

where K_e is coefficient dependent on ratio $D_i/2h_i$. Roche and Alix [39] also developed an empirical equation to estimate the plastic buckling pressure of internally pressurized ellipsoidal heads, as given by

$$P_b = K_c S_f\frac{t^{5/3}h_i^{4/3}}{(D_i/2)^3} \tag{1.22}$$

where $K_c = 88\left[1-2\left(\frac{2h_i}{D_i}\right)^2\right]^{-2/3}$, $S_f = (S_y + S_u)/2$. This empirical equation is based upon the experimental results of seventeen ellipsoidal heads fabricated by spinning. These heads are made of three kinds of materials: carbon steel, austenitic stainless steel and aluminium-magnesium alloy. These heads contain two types of radius-to-height ratios, i.e. 2.5 and 5, which are actually uncommon in engineering.

Section 1.3.3 is from Ref. [49] with permission of Elsevier.

1.4 Design of Ellipsoidal Heads

1.4.1 Design Standards

Figure 1.9 shows the evolution of design methods of ellipsoidal heads under internal pressure. In 1953, Boardman et al. [50] gave a historical overview of the work that led to the inclusion of design rules for ellipsoidal heads in 1950 ASME VIII code. Since the 1950s, ASME VIII-1 used a classical design formula based on the stress ratio factor (K) depending on radius-to-height ratio from the study of Maker in 1932 [1]. A similar formula based on the stress ratio factor (K) is also used in the Chinese standard GB/T 150.3 since the beginning of 1989 to the latest revision in 2011. In 1961, Shield and Drucker [51] developed a formula for calculating the limit pressure of internally pressurized torispherical heads. From 1970s to 2007, design curves based on the Shield & Drucker formula were used by ASME VIII-2 [52] to prevent against plastic collapse of torispherical and 2:1 ellipsoidal heads. These design curves have also been used in the Chinese industry standard JB 4732 [53] from the beginning of 1995 to the latest revision in 2005. Esztergar [54] in 1976 summarized the background information for design rules of vessel heads in 1970s ASME code. In 1986, Galletly [55, 56] developed a formula against buckling of torispherical heads. In 1991, Kalnins and Updike [57] developed design curves against yielding of head knuckle. In 1996 equivalent spherical shell formula based on membrane theory was recommended for preventing plastic collapse of head crown [58]. In 1999, Miller [59, 60] developed design formulas against buckling and plastic collapse of torispherical heads, where Miller also used the equivalent spherical shell formula to avoid plastic collapse of head crown. Since 2001, Miller design formulas

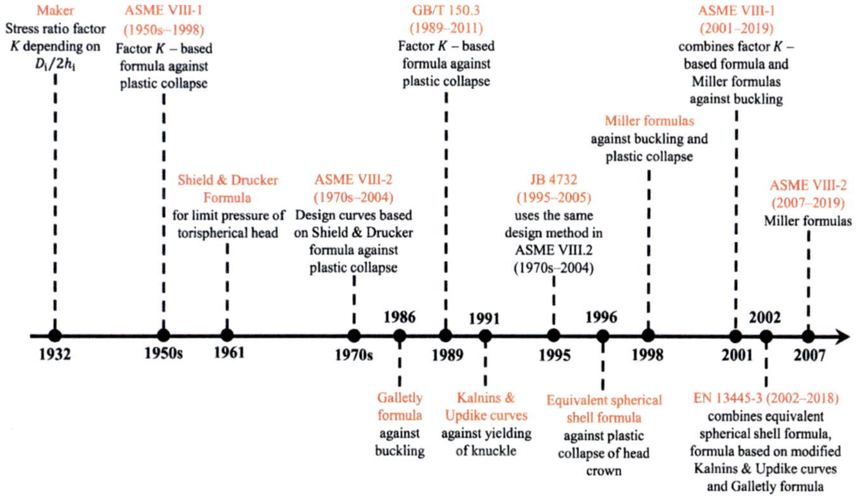

Fig. 1.9 Evolution of design methods of ellipsoidal heads under internal pressure

against buckling are used in ASME VIII-1 [61]; the design formula based on the stress ratio factor (K) is still used in ASME VIII-1 to prevent plastic collapse. From the beginning of 2002 to the latest revision of 2018, EN 13445-3 [62] combined the equivalent spherical shell formula against plastic collapse of head crown, formula against axisymmetric yielding of head knuckle based on modified Kalnins & Updike design curves [63] and Galletly formula against buckling to design the torispherical and ellipsoidal heads. Since 2007, ASME VIII-2 superseded ASME VIII-2 (before its 2004-revison) which was withdrawn, and has since used the Miller design formula to prevent buckling and plastic collapse of heads until now [64].

In addition, ASME Code Case 2260–2 [65] gives an alternative rule for design of ellipsoidal heads: the minimum required thickness of an ellipsoidal head with $D_i/2h_i \le 2.0$ shall be determined as an equivalent torispherical head using Formula (1.23), and a design Formula (1.24) for fatigue criterion is also provided in the code case [66, 67]. ASME Code Case N-284–4 [68] provides metal containment shell buckling design methods for ellipsoidal heads in class MC, TC, and SC construction, where the buckling capacity of shell is based on linear bifurcation buckling analyses reduced by capacity reduction factors which account for the effects of imperfections, geometric nonlinearity and boundary conditions and by plasticity reduction factors which account for nonlinearity in material properties [69].

$$t = \frac{M p_c L}{2\phi S - 0.2 p_c} \tag{1.23}$$

$$t = \frac{3 M_r p_c L E_r}{4 S_a} \tag{1.24}$$

where the values of M and M_r shall be obtained from Tables 1 and 2 of ASME Code Case 2260–2. The value of S_a shall be 793 MPa for all ferrous materials.

In ASME VIII-1 and ASME Code Case 2260–2, an acceptable approximation of a 2:1 ellipsoidal head is a torispherical head with a knuckle radius of 0.17 D_i and a crown radius of $0.9D_i$. Additionally, ellipsoidal heads with $t/L < 0.002$ in ASME VIII-1 are designed as equivalent torispherical heads to prevent failure of the knuckle, and the equivalent approach is presented in Table 1.2. It is seen that the values of r/D_i and L/D_i in Table 1.2 are determined from Formulas (1.10) and (1.11), as shown in Fig. 1.10. In ASME VIII-2 and EN 13445-3, ellipsoidal heads are designed as torispherical heads using Formulas (1.25) and (1.26) which are simple geometric equivalent approaches for ease of use in engineering. It is seen from Fig. 1.10 when $1.7 \le D_i/2h_i \le 2.2$, the values of r/D_i and L/D_i from Formulas (1.25) and (1.26)

Table 1.2 Geometric equivalency between ellipsoidal and torispherical heads in ASME VIII-1 [47]

$D_i/2h_i$	3.0	2.8	2.6	2.4	2.2	2.0	1.8	1.6	1.4	1.2	1.0
r/D_i	0.10	0.11	0.12	0.13	0.15	0.17	0.20	0.24	0.29	0.37	0.50
L/D_i	1.36	1.27	1.18	1.08	0.99	0.90	0.81	0.73	0.65	0.57	0.50

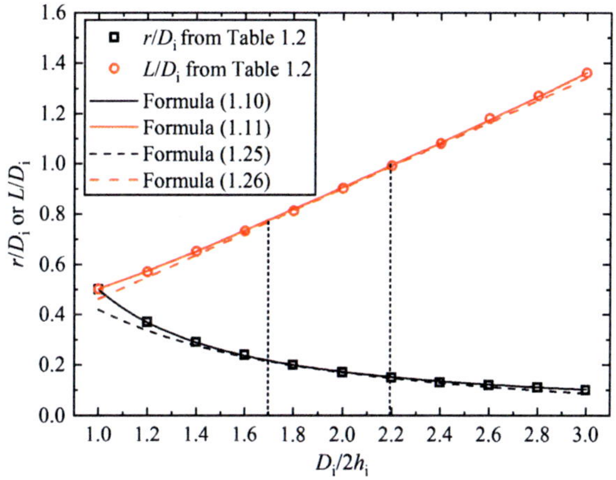

Fig. 1.10 Comparison of geometric equivalent approaches between ellipsoidal and torispherical heads (from Ref. [21] with permission of ASME)

are very close to Formulas (1.10) and (1.11), respectively. In fact, Formulas (1.25) and (1.26) are derived from Formulas (1.10) and (1.11). Recently, Seipp et al. [70] pointed out that these equivalent approaches are not supported by the shell theory method, and true ellipsoidal heads are being fabricated on basis of their survey; therefore, it is recommended that the rules for true ellipsoidal heads be developed rather than implementing these false equivalencies. The comparison of the geometric equivalent approaches can also be seen in our papers [21, 71].

$$r = D_{\mathrm{i}}[0.5/(D_{\mathrm{i}}/2h_{\mathrm{i}}) - 0.08] \tag{1.25}$$

$$L = D_{\mathrm{i}}[0.44(D_{\mathrm{i}}/2h_{\mathrm{i}}) + 0.02] \tag{1.26}$$

In 2007, the elastic–plastic analysis method was introduced to ASME VIII-2 and used in the design-by-analysis of pressure components. The effects of material strain hardening and geometric nonlinearity are considered in the elastic–plastic analysis method. However, the effects of both the material strain hardening and geometric nonlinearity are not included in current formulas for the design of ellipsoidal heads.

The detailed methods for design of ellipsoidal heads under internal pressure are summarized as follows and can also be seen in Ref. [71].

1.4.2 Protection Against Plastic Collapse

1.4.2.1 ASME VIII-1

ASME VIII-1 [47] uses classical design Formula (1.27) based on the stress ratio factor (K) given by Formula (1.14). Considering the factor (K) is the stress ratio of ellipsoidal head to the cylindrical shell, 2 K is the stress ratio of ellipsoidal head to spherical shell. Therefore, Formula (1.27) for ellipsoidal heads is accomplished by multiplying the calculated thickness of a spherical shell with diameter D_i by 2 K. This formula is also used in ASME III-1-NE [72].

$$t = \frac{K p_c D_i}{2\phi S - 0.2 p_c} \tag{1.27}$$

1.4.2.2 ASME VIII-2

In ASME VIII-2, ellipsoidal heads are designed as torispherical heads using geometric equivalent Formulas (1.25) and (1.26). The design method in ASME VIII-2 applies only to ellipsoidal heads for which $1.7 \leq D_i/2h_i \leq 2.2$. Other applicability scopes are as follows: $0.7 \leq L/D_i \leq 1.0$, $r/D_i \geq 0.06$ and $20 \leq L/t \leq 2000$.

Formula (1.28) is provided in ASME VIII-2 [8] to protect against bursting (plastic collapse) of torispherical and ellipsoidal heads subjected to internal pressure. The crown of torispherical head is equivalent to spherical shell with an inside radius of L. Using elastic shell theory, membrane stress (σ_m) of spherical shell is determined by force equilibrium (see Formula (1.29)), where L_m is mean radius ($L_m = L + 0.5t$). The burst pressure of spherical shells is given by Formula (1.30) which represents the condition when the membrane stress (σ_m) reaches ultimate tensile strength (S_u) of material [58, 67]. Substituting S_u, P_c and L_m with S, P and L, respectively, Formula (1.30) for predicting burst pressure becomes Formula (1.28) for preventing burst failure of heads.

$$P = \frac{2\phi S}{L/t + 0.5} \tag{1.28}$$

$$\sigma_m = \frac{p L_m}{2t} \tag{1.29}$$

$$P_c = \frac{2t S_u}{L_m} \tag{1.30}$$

1.4.2.3 GB/T 150.3

The Chinese standard GB/T 150.3 [48] uses Formula (1.31) to determine the required thickness of ellipsoidal heads under internal pressure. This formula is similar to ASME VIII-1; the design formula in GB/T 150.3 is based on membrane theory, while the $0.2p_c$ modifier in ASME VIII-1 is an adjustment to account for the increase in stress level along the wall in thick heads [73].

$$t = \frac{Kp_c D_i}{2\varphi S - 0.5p_c} \tag{1.31}$$

1.4.2.4 JB 4732

In Chinese industry standard JB 4732 [53], 2:1 ellipsoidal heads are designed as equivalent torispherical heads with $r/D_i = 0.17$ and $L/D_i = 0.9$, which satisfies the equivalent Formulas (1.25) and (1.26). However, for ellipsoidal heads which have $D_i/2h_i$ values different from 2, JB 4732 does not give the detailed specification on the equivalency.

For the intermediate thickness heads with $t/L \geq 0.002$ up to a t/L where $p_c/S \leq 0.08$ (approximately $t/L = 0.04$ to 0.05), the minimum required thickness shall be determined by using the design curves in Fig. 1.11. Interpolation may be used for r/D_i values which fall within the range ($0.06 \leq r/D_i \leq 0.2$) of the curves; however, no extrapolation of the curves is permitted.

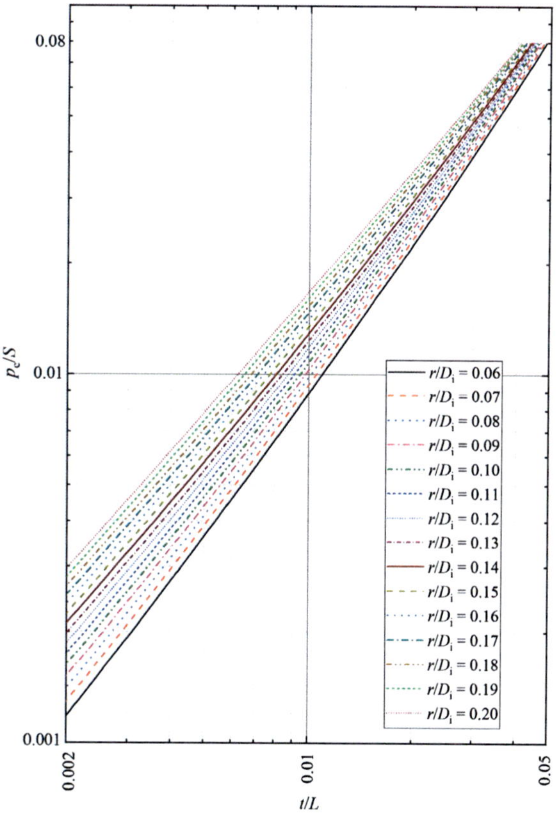

Fig. 1.11 Design curves for torispherical and 2:1 ellipsoidal heads [52]

For designs where $p_c/S > 0.08$, which is above the upper limit of Fig. 1.11, the head thickness shall be determined by Formula (1.32):

$$t = \frac{D_i}{2}\left[\exp(p_c/S) - 1\right] \tag{1.32}$$

For thin heads with $t/L < 0.002$, which is below the lower limit of Fig. 1.11, the head design must be based on stress analysis, experimental stress analysis or fatigue analysis. But there is no formula for the design of thin heads.

The design curves in Fig. 1.11 for heads of intermediate thickness are determined by Formula (1.33) for natural logarithms or Formula (1.34) for common base logarithms [52].

$$\begin{aligned}\ln(t/L) = &-1.26176643 - 4.5524592\frac{r}{D_i} + 28.933179\left(\frac{r}{D_i}\right)^2 \\ &+ \left[0.66298796 - 2.2470836\frac{r}{D_i} + 15.682985\left(\frac{r}{D_i}\right)^2\right]\ln(p_c/S) \\ &+ \left[0.26878909 \times 10^{-4} - 0.42262179\frac{r}{D_i} + 1.8878333\left(\frac{r}{D_i}\right)^2\right][\ln(p_c/S)]^2\end{aligned} \tag{1.33}$$

$$\begin{aligned}\log(t/L) = &-0.5479782 - 1.9771079\frac{r}{D_i} + 12.565520\left(\frac{r}{D_i}\right)^2 \\ &+ \left[0.66298796 - 2.2470836\frac{r}{D_i} + 15.682985\left(\frac{r}{D_i}\right)^2\right]\log(p_c/S) \\ &+ \left[0.61890975 \times 10^{-4} - 0.97312263\frac{r}{D_i} + 4.3468967\left(\frac{r}{D_i}\right)^2\right][\log(p_c/S)]^2\end{aligned} \tag{1.34}$$

Formulas (1.33) and (1.34) (i.e., design curves in Fig. 1.11) are determined on basis of the Formula (1.35) for calculating the limit pressure of internally pressurized torispherical heads from the study of Shield and Drucker [51], as mentioned by Esztergar [54]. It is easy to calculate pressure by using Formula (1.35), but there is a difficulty of calculation for head thickness; thus, Formula (1.35) becomes the design Formulas (1.33) or (1.34) for ease of calculating the head thickness.

$$\frac{P_L}{S_y} = \left(0.33 + 5.5\frac{r}{D_i}\right)\frac{t}{L} + 28\left(1 - 2.2\frac{r}{D_i}\right)\left(\frac{t}{L}\right)^2 - 0.0006 \tag{1.35}$$

For thin heads with $t/L < 0.002$, buckling is the controlling failure mode, however, which is not considered in this method.

For thick heads with $p_c/S > 0.08$ (approximately $t/L = 0.04$ to 0.05), the design Formula (1.32) for cylindrical shells based on limit analysis is used to prevent collapse of heads due to conservatism of the cylindrical shell formula [74].

1.4.2.5 EN 13445-3

In EN 13445-3, ellipsoidal heads are designed as torispherical heads using geometric equivalent Formulas (1.25) and (1.26). The design method in EN 13445-3 applies only to ellipsoidal heads for which $1.7 < D_i/2h_i < 2.2$. Other applicability scopes are as follows: $L/D_i \leq 1.0$, $0.06 \leq r/D_i \leq 0.2$ and $12.5 \leq D_o/t \leq 1000$.

EN 13445-3 [62] uses Formula (1.36) to limit membrane stress in head crown in order to prevent plastic collapse of head crown. Formula (1.36) to calculate required thickness is similar to ASME VIII-2 which provides Formula (1.28) for calculating allowable pressure to prevent failure of head crown.

$$t = \frac{p_c L}{2\phi S - 0.5 p_c} \tag{1.36}$$

Additionally, EN 13445-3 [62] adopts Formula (1.37) to avoid axisymmetric yielding of head knuckle. The parameter β_t is determined from Fig. 1.12 or the Eqs. (7.5–9)–(7.5–17) in EN 13445. Formula (1.38) for calculating allowable pressure to prevent axisymmetric yielding of head knuckle is also provided in EN 13445-3. The method is based on the work of Kalnins and Updike [57, 63]. The failure criterion chosen by Kalnins & Updike is the twice elastic slope criterion applied to the deformation of the head apex. The theoretical model includes shape change, but material strain hardening is not considered.

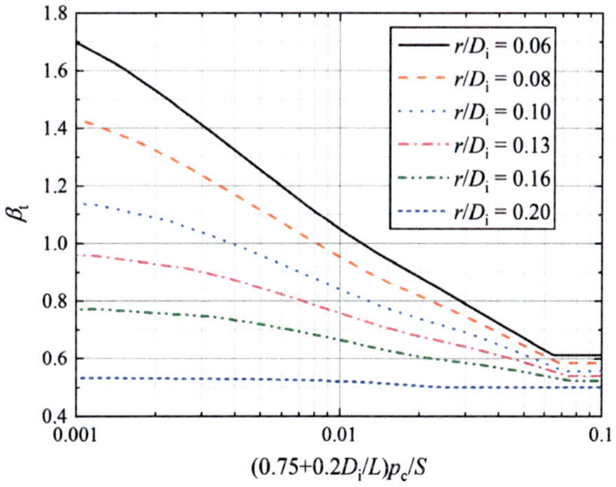

Fig. 1.12 Parameter β_t for design of heads under internal pressure (EN13445 [62])

$$t_y = \frac{\beta_t p_c (0.75L + 0.2D_i)}{S} \tag{1.37}$$

$$P = \frac{St}{\beta_t (0.75L + 0.2D_i)} \tag{1.38}$$

1.4.3 Protection Against Local Buckling

ASME VIII-1 and VIII-2, and EN 13445-3 provide the design formulas shown below for protection against local buckling of ellipsoidal heads under internal pressure. It is noted that, however, Chinese standards GB/T 150.3 and JB 4732 do not give design formulas against local buckling. In GB/T 150.3, for $D_i/2h_i \leq 2.0$, local buckling is not considered in head design when $D_i/t \leq 666$; while for $D_i/2h_i > 2.0$, local buckling is not considered when $D_i/t \leq 333$.

1.4.3.1 ASME VIII-1 and VIII-2

In ASME VIII-1 [47] and VIII-2 [8], the minimum required thickness to prevent buckling of ellipsoidal heads under internal pressure shall be calculated using the following procedure.

(a) Determine the inside diameter (D_i) and radius-to-height ratio ($D_i/2h_i$), and determine the knuckle radius (r) and the crown radius (L) using Formulas (1.25) and (1.26) respectively, and assume value for the wall thickness (t).
(b) Compute L/D_i, r/D_i, and L/t ratios and determine if the following scopes are satisfied: $0.7 \leq L/D_i \leq 1.0$, $r/D_i \geq 0.06$ and $20 \leq L/t \leq 2000$. If the scopes are satisfied, then proceed to Step (c); otherwise, the head shall be designed in accordance with design-by-analysis methods.
(c) Calculate the following geometric constants:

$$\beta_{th} = \arccos\left(\frac{0.5D_i - r}{L - r}\right) \tag{1.39}$$

$$\phi_{th} = \frac{\sqrt{Lt}}{r} \tag{1.40}$$

$$R_e = \begin{cases} (0.5D_i - r)/[\cos(\beta_{th} - \phi_{th})] + r, \phi_{th} < \beta_{th} \\ 0.5D_i, \phi_{th} \geq \beta_{th} \end{cases} \tag{1.41}$$

(d) Compute the coefficients C_1 and C_2 using the following equations.

$$C_1 = \begin{cases} 9.31r/D_\mathrm{i} - 0.086, & r/D_\mathrm{i} \le 0.08 \\ 0.692r/D_\mathrm{i} + 0.605, & r/D_\mathrm{i} > 0.08 \end{cases} \tag{1.42}$$

$$C_2 = \begin{cases} 1.25, & r/D_\mathrm{i} \le 0.08 \\ 1.46 - 2.6r/D_\mathrm{i}, & r/D_\mathrm{i} > 0.08 \end{cases} \tag{1.43}$$

(e) Calculate the value of internal pressure expected to produce elastic buckling of the knuckle.

$$S_\mathrm{e} = C_1 E(t/r) \tag{1.44}$$

$$P_\mathrm{e} = \frac{S_\mathrm{e} t}{C_2 R_\mathrm{e}[(0.5R_\mathrm{e}/r) - 1]} \tag{1.45}$$

(f) Calculate the value of internal pressure that will result in a maximum stress in the knuckle equal to the material yield strength.

$$P_\mathrm{y} = \frac{S_\mathrm{y} t}{C_2 R_\mathrm{e}[(0.5R_\mathrm{e}/r) - 1]} \tag{1.46}$$

(g) Calculate the value of internal pressure expected to result in buckling failure of the knuckle. Formula (1.47) is used in ASME VIII-1. When $G > 1.0$, ASME VIII-2 uses Formula (1.48) which is very close to Formula (1.47) [71, 75].

$$P_\mathrm{f} = \begin{cases} 0.6P_\mathrm{e}, & G \le 1.0 \\ 0.408P_\mathrm{y} + 0.192P_\mathrm{e}, & 1.0 < G \le 8.29 \\ 2.0P_\mathrm{y}, & G > 8.29 \end{cases} \tag{1.47}$$

$$P_\mathrm{f} = \left(\frac{0.77508G - 0.20354G^2 + 0.019274G^3}{1 + 0.19014G - 0.089534G^2 + 0.0093965G^3}\right)P_\mathrm{y}, \quad G > 1.0 \tag{1.48}$$

where $G = P_\mathrm{e}/P_\mathrm{y}$.

(h) Calculate the allowable pressure based on a buckling failure of the knuckle.

$$P_k = P_f/1.5 \tag{1.49}$$

The above method in ASME VIII-1 and VIII-2 for preventing buckling of ellipsoidal heads under internal pressure is based on the formulas of Miller [59, 60]. Miller formulas were developed for preventing the buckling and burst failures of torispherical heads. Formula (1.28) for preventing burst is determined from Formula (1.30) for burst pressure of spherical shell; see Sect. 1.4.2.2. Formulas (1.39)–(1.49) for buckling are derived by applying reduction factors to theoretical elastic buckling stress. The reduction factors which consider the effects of geometric imperfections, nonlinearity material properties and residual stresses were derived from test data.

1.4.3.2 EN 13445-3

In EN 13445-3 [62], the required thickness to prevent knuckle buckling of internally pressurized ellipsoidal heads is determined by Formula (1.50). The knuckle radius (r) and the crown radius (L) are determined using Formulas (1.25) and (1.26), respectively. It is not necessary to calculate t if $t_y > 0.005D_i$, where t_y is the required thickness determined by Formula (1.37) to avoid axisymmetric yielding of head knuckle.

$$t = (0.75L + 0.2D_i)\left[\frac{p_c}{111S}\left(\frac{D_i}{r}\right)^{0.825}\right]^{\left(\frac{1}{1.5}\right)} \tag{1.50}$$

where, $S = R_{p0.2}/1.5$, except for cold spun seamless austenitic stainless steel, where: $S = 1.6R_{p0.2}/1.5$. The 1.6 factor for cold spun heads takes account of work hardening during spinning.

Galletly [55, 56] concluded with a suggested design Formula (1.51), which was modified slightly to Formula (1.50) for use in EN 13445-3 to prevent buckling of ellipsoidal heads under internal pressure. The basis of Formula (1.51) is a curve fit to the results of many large deflection plastic buckling calculations, with a knock-down factor based on the lower bound of a limited number of test results.

$$P = \frac{80\gamma S_y (r/D_i)^{0.825}}{(D_i/t)^{1.5}(L/D_i)^{1.15}} \tag{1.51}$$

where $\gamma = 1.0$ for steel heads fabricated by segments; $\gamma = 1.6$ for cold spun steel heads.

1.5 Fabrication of Ellipsoidal Heads

1.5.1 Fabrication Methods

The following methods are usually used to construct ellipsoidal heads: pressing, spinning and assembly from formed segments.

1.5.1.1 Pressing

For ellipsoidal heads formed by pressing, a circular flat plate (or welded plate) is directly pressed to the required shape in a concave die, as shown in Fig. 1.13. Cold pressing is a forming process at room temperature. Warm pressing is a process of deforming metal heated to an intermediate temperature (above room temperature, but far below the recrystallization temperature) [76]. Hot pressing is a process which is performed at temperatures above the recrystallization temperature. In order to avoid the associated microstructural changes caused by heat assisted plastic deformation, some material such as aluminum sheets may be deformed at cryogenic temperatures, i.e., cryogenic forming [77].

1.5.1.2 Spinning

For ellipsoidal heads formed by spinning, a circular flat plate (or welded plate) is first pressed to the crown radius with a die. The knuckle of the head is then formed

Fig. 1.13 Head pressing process

Fig. 1.14 Head spinning process

on the edge by turning-in the rim with rollers as the head is rotated about its axis, as shown in Fig. 1.14. Similar to pressing process, head spinning fabrication includes cold, warm and hot spinning.

1.5.1.3 Assembly from Formed Segments

For ellipsoidal heads assembled from formed segments, the crown of the head is made from one plate or welded by several plates, and formed to the crown radius. The knuckle of the head is welded from several segment which are first formed by pressing it with a die. Finally, the crown and the knuckle are welded together to assemble an ellipsoidal head, as shown in Fig. 1.15.

1.5.2 Quality Requirements

1.5.2.1 Shape Deviation

The requirements for shape deviation of an ellipsoidal head in the pressure vessel codes ASME VIII-1 and VIII-2, Chinese standards GB/T 150.4 [78], GB/T 25198 [79] and JB 4732 are as follows: The inner surface of the head shall not deviate outside of the specified shape by more than 1.25% of D_n nor inside the specified shape by more than 0.625% of D_n, where D_n is the nominal inside diameter of the head. Such deviations shall be measured perpendicular to the specified shape and shall not be abrupt. The knuckle radius shall not be less than that specified. Measurements for determining the above deviations shall be taken from the surface of the base metal and not from weld joints.

Fig. 1.15 Assembly from formed segments

EN 13445-4 [80] has different requirements for shape deviation: head crown radius shall not be greater than that specified in the design, and head knuckle radius shall not be less than that specified.

1.5.2.2 Forming Strain

Strain occurs in head forming, which may unduly impair the mechanical properties of material. Thus, in pressure vessel codes and standards, post-forming heat treatment is required when the forming strain exceeds limitation. The formulas for calculating forming strain and limitations of forming strain are shown below.

(1) Formulas for calculating forming strain

(a) ASME VIII-1

In ASME VIII-1 [47], Formula (1.52) is used for calculation of forming strain for heads, which represents the maximum engineering strain of the metal fiber in the meridional direction due to bending in head forming.

$$\varepsilon_f = \frac{75t}{R_f}\left(1 - \frac{R_f}{R_{om}}\right) \tag{1.52}$$

where R_{om} represents original mean radius, equal to infinity for a flat plate; R_f represents the final mean radius after forming and is equal to $0.17D_n$ for 2:1 ellipsoidal heads.

(b) ASME VIII-2

In ASME VIII-2 [8], for all one-piece, double-curved circumferential products, which are formed by processes such as dishing or cold spinning (for example, dished heads or cold spun heads), Formula (1.53) is used for calculating forming strain which represents the maximum true circumferential compressive membrane strain of the head edge area as a result of the reduction of the diameter of the circular blank to the outside diameter of the head.

$$\varepsilon_{\mathrm{f}} = 100\ln\left(\frac{D_{\mathrm{b}}}{D_{\mathrm{f}} - 2t}\right) \tag{1.53}$$

where D_{b} is the diameter of a flat plate or an intermediate product; D_{f} is the final outside diameter of heads after forming.

For heads assembled from formed segments (for example, dished segments of ellipsoidal heads), Formula (1.52) is used for calculation of forming strain for this type of heads.

(c) EN13445-4

In EN13445-4 [80], Formula (1.53) is used for calculation of the forming strain for dished heads.

Whereas, the following Formula (1.54) shall be used when calculating forming strain of heads assembled from segments.

$$\varepsilon_{\mathrm{f}} = \frac{100t}{R_{\mathrm{sm}}} \tag{1.54}$$

where R_{sm} is the smallest mean radius of the segment (calculated by using the mean crown radius and mean knuckle radius of an ellipsoidal head).

(d) GB/T 150.4 and GB/T 25198

Formula (1.52) is also used for calculating forming strain of heads in the Chinese standards GB/T 150.4 [79] and GB/T 25198 [79].

(2) Limitations of forming strain

(a) ASME VIII-1 and ASME VIII-2

In ASME VIII-1 and ASME VIII-2, the forming strain limitation is 5% for all carbon and low alloy steels except P-No. 1, Group Nos. 1 and 2 steels, as shown in Paragraph UCS of ASME VIII-1 and Paragraph 6.1.2.3 of ASME VIII-2. For pressure vessels and vessel parts that are constructed of high alloy steel, the forming strain limitations are shown in Table UHA-44 of ASME VIII-1 and in Table 6.2.B of ASME VIII-2:

For example with the 304 material, the forming strain limitation is 20% in design temperature lower than or equal to 675 °C, and the limitation is reduced to 10% in design temperature exceeding 675 °C.

(b) EN 13445-4

The forming strain limitation is provided in Paragraph 9.4 of EN13445-4. For flat products such as heads, the forming strain limitation is 5%.

(c) GB/T 150.4

In GB/T 150.4, the forming strain of vessels or components made of carbon and low alloy steels and other material is limited to 5%. For austenite stainless steel, the forming strain is limited to 15%, however it is limited to 10% when design temperature is lower than −100 °C or greater than 675 °C.

1.6 Special Ellipsoidal Heads

1.6.1 Heads Under External Pressure

External pressure is defined as pressure acting on the convex side of the shell. Ellipsoidal heads, used in vacuum, jacketed and submarine vessels, are subjected to external pressure.

ASME VIII-1 provides the design methods for externally pressurized ellipsoidal heads using external pressure charts; see Paragraph UG-33(d) of ASME VIII-1. Aside from ASME VIII-1, GB/T 150.3 and JB 4732 also use similar external pressure charts to design ellipsoidal heads, as provided in Paragraph 5.3.3 of GB/T 150.3 and in Paragraph 8.4.2 of JB 4732, respectively. Alternative design rules for ellipsoidal heads using formulas are provided in Paragraph 4.4.9 of ASME VIII-2. In addition, EN1 3445-3 combines formulas and curves to design ellipsoidal heads under external pressure; see Paragraph 8.8.3 of EN 13445-3.

Ellipsoidal heads under external pressure are designed as spherical shells. In ASME VIII-1, GB/T 150.3 and JB 4732, the equivalent outside spherical radius $R_o = K_o D_o$, where spherical radius factor (K_o) depends on the ellipsoidal head proportion ($D_o/2h_o$) and is given in Table 1.3. In ASME VIII-2 [8], K_o is given by Formula (1.55). In fact, the values of K_o in Table 1.3 is equal to the values determined by Formula (1.55). In EN 13445-3 [62], an ellipsoidal head subjected to

Table 1.3 Spherical radius factor (K_o) for externally pressurized ellipsoidal heads [47]

$D_o/2h_o$	1.0	1.2	1.4	1.6	1.8	2.0	2.2	2.4	2.6	2.8	3.0
K_o	0.5	0.57	0.65	0.73	0.81	0.9	0.99	1.08	1.18	1.27	1.36

Note Interpolation permitted for intermediate values

external pressure shall be designed as a spherical shell of mean radius R_m equal to the maximum radius of the crown, as given by Formula (1.56).

$$K_o = 0.25346 + 0.13995\left(\frac{D_o}{2h_o}\right) + 0.12238\left(\frac{D_o}{2h_o}\right)^2 - 0.015297\left(\frac{D_o}{2h_o}\right)^3 \quad (1.55)$$

$$R_m = D_o^2/(4h_o) \quad (1.56)$$

1.6.2 Heads with Variable Thicknesses

Ellipsoidal heads with variable thicknesses are used in engineering due to their good mechanical performance and light weight. There are two types of ellipsoidal heads with variable thicknesses: (a) For large heads assembled by segments, they have different crown and knuckle thicknesses; and (b) The thickness of the head is continuously changing.

1.6.2.1 Heads with Different Crown and Knuckle Thicknesses

ASME VIII-2 and EN 13445-3 provide the methods for design of ellipsoidal heads with different crown and knuckle thicknesses, as shown in Fig. 1.16; it is permissible to reduce the thickness of the head crown smaller than the knuckle thickness.

(1) ASME VIII-2

In ASME VIII-2 [8], the minimum required thickness of the head crown shall be determined in accordance with the design methods for spherical heads using Formula (1.57). The diameter of the circular area corresponding to the head crown shall be less than or equal to $0.8D_i$ (Fig. 1.16).

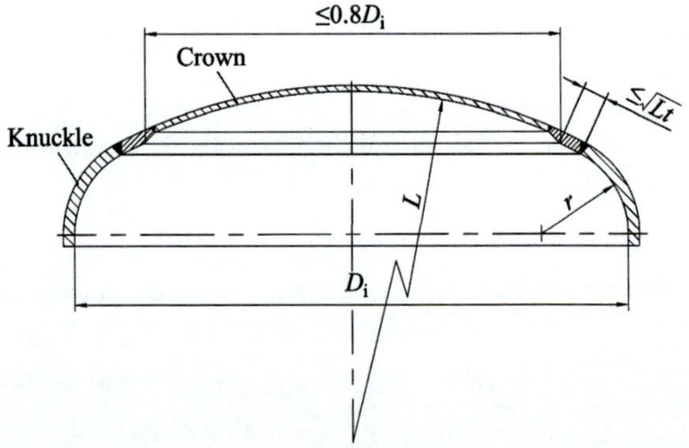

Fig. 1.16 Heads with different crown and knuckle thicknesses

$$t = \frac{D_i}{2}\left[\exp\left(\frac{p_c}{2S}\right) - 1\right] \tag{1.57}$$

The minimum required thickness of the head knuckle shall be determined in accordance with the design methods of ASME VIII-2 presented in Sects. 1.4.2 and 1.4.3. The transition in thickness shall be located on the inside surface of the thicker part, and shall have a taper not exceeding 1:3.

(2) EN 13445-3

In EN 13445-3 [62], it is permissible to reduce the thickness of the head crown to the value t_c over a circular area that shall not come closer to the knuckle than the distance $\sqrt{Lt}$ (Fig. 1.16), where t_c is the required thickness to limit membrane stress in head crown using Formula (1.36), and t is the maximum value between the required thicknesses to prevent axisymmetric yielding and local buckling of the head knuckle using Formulas (1.37) and (1.50).

1.6.2.2 Heads with Continuously Changing Thicknesses

Based on Formulas (1.1) and (1.2) for the meridional stress (σ_φ) and circumferential stress (σ_θ) of shell of revolution, a formula is derived for calculating von Mises equivalent stress in the dished head under uniform internal pressure [24], as given by

$$\sigma_{eq} = \sqrt{\sigma_\theta^2 - \sigma_\theta\sigma_\varphi + \sigma_\varphi^2} = \frac{pR_2}{2t}\sqrt{3\left(1 - \frac{R_2}{R_1}\right) + \left(\frac{R_2}{R_1}\right)^2} \tag{1.58}$$

where R_1 and R_2 are the principal radii of ellipsoidal head curvature, and determined using Formulas (1.3) and (1.4).

Based on Formula (1.58) for equivalent stress in the head, a design Formula (1.59) can be developed for determining the required thickness with respect to the strength of ellipsoidal heads when equivalent stress is equal to allowable stress of material (i.e., $\sigma_{eq} = S$). Since R_1 and R_2 are the functions of the meridian length, the required thickness given by Formula (1.59) is changing with the meridian length. Therefore, Formula (1.59) is used to design iso-strength ellipsoidal heads with continuously changing thicknesses.

$$t = \frac{p_c R_2}{2S}\sqrt{3\left(1 - \frac{R_2}{R_1}\right) + \left(\frac{R_2}{R_1}\right)^2} \tag{1.59}$$

1.6.3 Heads with Nozzles

Ellipsoidal heads with nozzles shall be adequately reinforced in the area adjacent to the nozzle opening. This is to compensate for the reduction of the pressure bearing section. The reinforcement shall be obtained by one of the following methods [62]: (a) increasing the wall thickness of the head; (b) using a reinforcing ring; (c) increasing the wall thickness of the nozzle; and (d) using a combination of the above.

Chapter 2
Buckling of Ellipsoidal Heads

2.1 Concept of Buckling

As stated in the book by Eslami [1], a structure is capable of withstanding the induced stresses and remaining in its original stable equilibrium condition to the point that a small change in applied loads does not produce a large deformation which leads to divergence from the equilibrium condition. Or, if the structure deviates or diverges from its equilibrium position due to the application of a small change in the applied loads, the equilibrium condition becomes unstable. Buckling is a kind of instability failure in which a structure loses its configuration due to compressive stresses. To better understand buckling, the usual buckling types are presented as follows.

Two buckling phenomena named by Bushnell [2, 3] are (1) snap-through buckling or limit point buckling, characterized by collapse at the maximum point in a load–displacement curve, and (2) bifurcation buckling. These two types of buckling are illustrated in Fig. 2.1 in the case of a rather thick axially compressed cylinder. The cylinder deforms approximately axisymmetrically along the equilibrium path OA until a maximum or collapse load is reached at point A, following either path ABF along which it continues to deform axisymmetrically or some other path ABE along which it first deforms axisymmetrically from A to B, then non-axisymmetrically from B to E. Snap-through buckling or limit point buckling occurs at point A and bifurcation buckling at point B.

Buckling occurs when material is in an elastic state, which is called elastic buckling. Elastic buckling pressure mainly depends on geometric parameters, Young's modulus and Poisson's ratio. Shells with relatively smaller diameter-to-thickness ratio are prone to buckle in the elastic–plastic or plastic region of the material. This type of buckling is called elastic–plastic or plastic buckling. Plastic buckling pressure mainly depends on geometric parameters, yield strength and material hardening.

Some shells collapse by global buckling, which always leads to the complete loss of load-carrying capacity, such as externally pressurized cylinder that failed by overall buckling shown in Fig. 2.2a. For some shells such as internally pressurized

J. Zheng and K. Li, *New Theory and Design of Ellipsoidal Heads for Pressure Vessels*,
https://doi.org/10.1007/978-981-16-0467-6_2

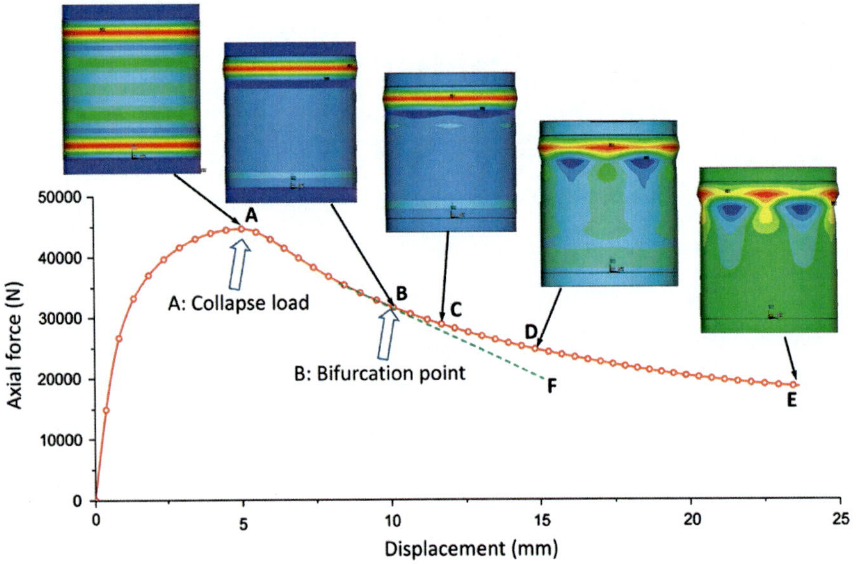

Fig. 2.1 Buckling behavior of rather thick cylinder under axial compression

Fig. 2.2 **a** Global buckling of externally pressurized cylinder; **b** Local buckling of internally pressurized ellipsoidal head

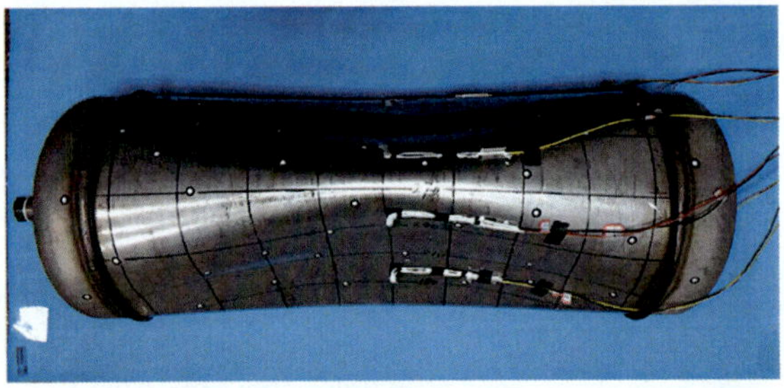

(a)

(b)

ellipsoidal heads, buckling occurs at local positions, as shown in Fig. 2.2b. The shells failed by local buckling can sometimes carry further load.

In addition, there are other types of buckling [4]: thermal buckling, creep buckling, dynamic buckling, etc. These buckling types are not further discussed in this book.

Chapter 2 is intended to provide an in-depth investigation on buckling of ellipsoidal heads under internal pressure, and adapted from Refs. [5–10] with permission of ASME and Elsevier.

2.2 Key Technologies for Buckling Test of Large (Φ5000) Ellipsoidal Heads

2.2.1 Design of Reusable Test Vessel with Large Ellipsoidal Heads

2.2.1.1 Large Ellipsoidal Heads

We performed buckling experiments on six large, thin ellipsoidal heads, as shown in Fig. 2.3. Table 2.1 lists the geometric parameters of the tested ellipsoidal heads. These heads—ZJU-FA, ZJU-FT, ZJU-FN, each in duplicate—were fabricated in three radius-to-height ratios ($D_i/2h_i$): ZJU-FA heads: 1.728; ZJU-FT heads: 2.0; and ZJU-FN heads: 1.591. The nominal diameter (D_i) of the ZJU-FA heads was 4797 mm, while ZJU-FT and ZJU-FN heads each had a nominal diameter of 5000 mm. All

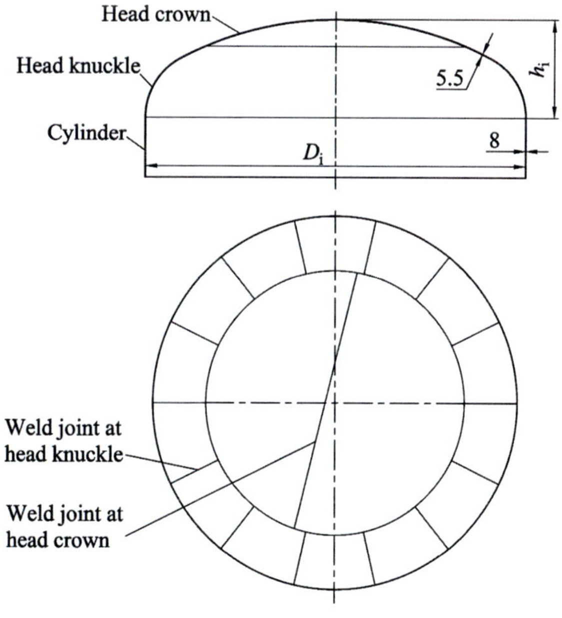

Fig. 2.3 Schematic diagram of tested large ellipsoidal heads

Table 2.1 Geometric parameters of tested large ellipsoidal heads

Head No	Nominal diameter (mm)	Nominal thickness (mm)	Radius-to-height ratio
ZJU-FA	4797	5.5	1.728
ZJU-FT	5000	5.5	2.0
ZJU-FN	5000	5.5	1.591

Table 2.2 Measured mechanical properties of SA738 Gr. B steel (from Ref. [10])

Specimen No	Head No	Young's modulus (MPa)	Yield strength (MPa)	Tensile strength (MPa)
S1	ZJU-FA1	208,336.5	509	602
S2	ZJU-FA2	186,290.0	588	662
S3	ZJU-FT1	199,694.6	612	674
S4	ZJU-FT2	201,444.0	558	626
S5	ZJU-FN1	197,087.4	609	681
S6	ZJU-FN2	199,987.1	615	673

heads had a nominal thickness of 5.5 mm. Each head was attached to a cylinder with a nominal thickness of 8 mm. Due to their large diameters, all heads were assembled by welding formed segments (Fig. 2.3).

The tested ellipsoidal heads were made from ASME SA738 Gr. B steel. Before the fabrication of the heads, tensile test specimens were cut from the SA738 Gr. B plates, and tensile tests were performed according to the ASTM A370 code [11]. The measured mechanical properties of these specimens are given in Table 2.2, and their corresponding true stress-true strain curves are shown in Fig. 2.4.

2.2.1.2 Test Vessel

As shown in Fig. 2.5, the tested heads with cylinders were welded to a test device in order to assemble a test vessel for internally pressurized buckling experiments. The ZJU-FA heads were tested first by connecting them to the test device via a transitional section consisting of a cylindrical shell with T-type reinforced ring and conical shell (Fig. 2.5a). The conical shell was used for test heads with different diameters. After testing the ZJU-FA heads with a diameter of 4797 mm, the ruptured ZJU-FA2 head along with the conical shell section was removed from the A-A position in Fig. 2.5a, and the test device was reused for testing the ZJU-FN and ZJU-FT heads with a diameter of 5000 mm. After testing a ZJU-FN head, Cylinder B was cut at section B-B (Fig. 2.5b) and a ZJU-FT head (Fig. 2.5c) was then welded to Cylinder B.

The design pressure of the test vessel is defined as the plastic collapse pressure of the tested ellipsoidal heads. This provides a substantial margin of safety which permits pressurizing the ellipsoidal head during the test until its rupture. The plastic

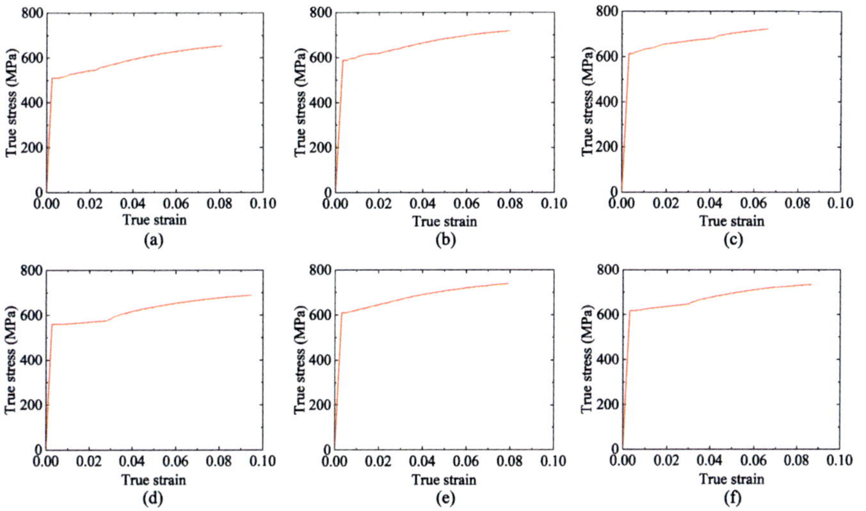

Fig. 2.4 Measured true stress-true strain curves of SA738 Gr. B from tensile specimens Nos. **a** S1; **b** S2; **c** S3; **d** S4; **e** S5; **f** S6 (from Ref. [10])

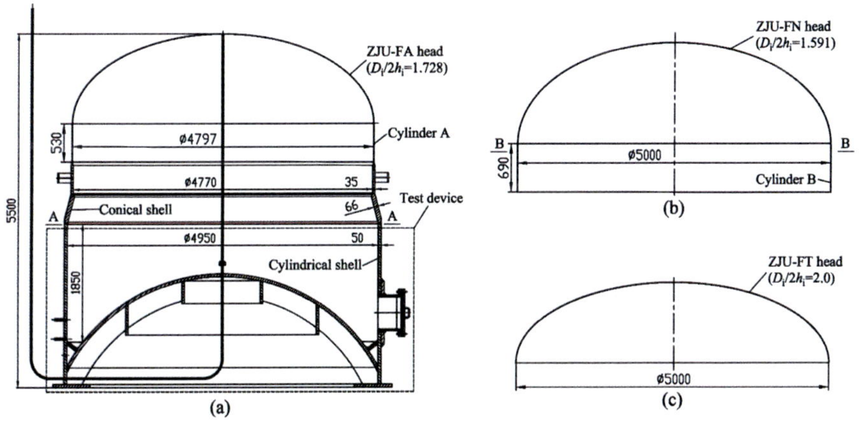

Fig. 2.5 Schematic diagram of test vessel (mm): **a** Test vessel with ZJU-FA head; **b** ZJU-FN head; **c** ZJU-FT head (from Refs. [7, 9] with permission of ASME and Elsevier)

collapse pressure of the ellipsoidal head subjected to internal pressure was determined using the finite element method presented by Liu et al. [12] where the arc-length algorithm and restart analysis were used, and the effects of the elastic–plastic material model and geometric nonlinearity were considered. The true stress-true strain curve of SA738 Gr. B from ASME VIII-2 (Sect. 1.1.2) was used for the material model (Fig. 2.6a), in which the yield strength is 415 MPa and the maximum specified tensile

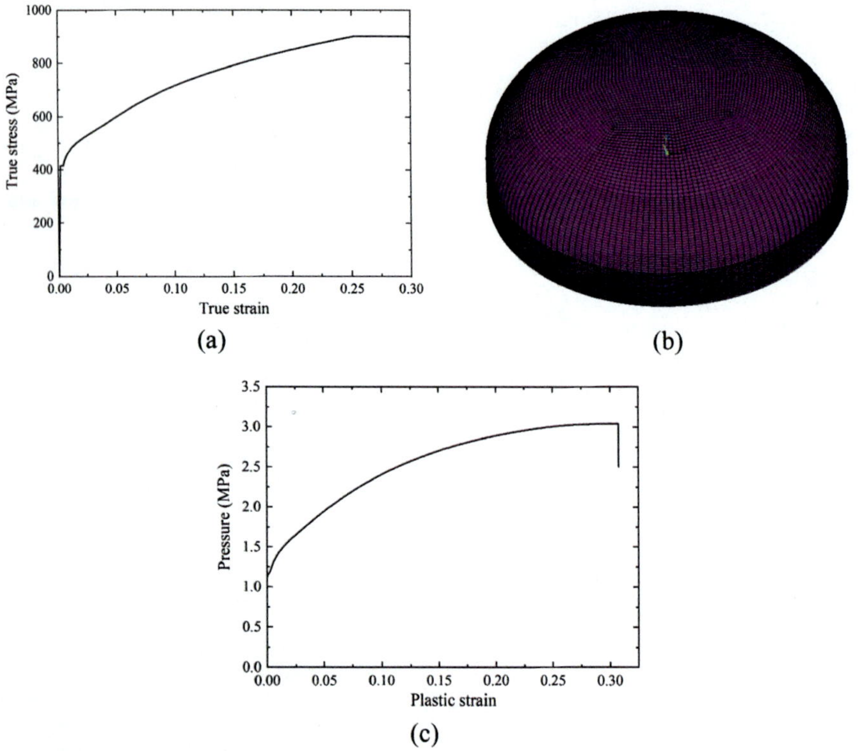

Fig. 2.6 Prediction of plastic collapse pressure of ellipsoidal head: **a** True stress-true strain curve; **b** Mesh model; **c** Pressure-plastic strain curve (from Ref. [7] with permission of ASME)

strength is 705 MPa according to ASME II-D [13]. A geometric model for the ZJU-FA heads was created using the nominal values of the dimensions shown in Table 2.1. A mesh model using the 4-node structural shell element SHELL181 is shown in Fig. 2.6b. Pressure was applied to the internal surface of the ellipsoidal head, and the bottom edge of the cylinder of the ellipsoidal head was fixed to eliminate the rigid body displacement of the model. Figure 2.6c shows the pressure-von Mises plastic strain curve at the apex of the ellipsoidal head. As the strain increases, the pressure initially increases, then decreases when the strain becomes greater than 0.308. After the curve enters the descending stage, the structure starts to collapse in the form of unstable plastic tension. The curve in Fig. 2.6c shows that the plastic collapse pressure of the ellipsoidal head is 3.0 MPa. Therefore, the design pressure (P_d) of the test vessel is 3.0 MPa.

A concave shallow spherical head was designed to reduce production cost compared to that of a flat head design, and a support for test vessels was omitted. As the spherical head was subjected to external pressure, it was reinforced by rings and stiffeners in order to avoid its buckling, as shown in Figs. 2.5a and 2.7. The spherical head under external pressure needed to be evaluated to protect it against instability with regard to the Type-1 buckling criteria outlined in Part 5.4 of ASME

Fig. 2.7 Spherical head reinforced by rings and stiffeners (from Ref. [10])

VIII-2. A geometric model of the spherical head was created using the dimensions shown in Fig. 2.5a. The eight-node solid element SOLID45 was used to generate the mesh of the model, as shown in Fig. 2.8a. The load used for the analysis was the design pressure of 3.0 MPa, which was applied to the pressure boundary of the model. Symmetry constraints were applied to the cut edges of the model, and the upper side of the cylinder was fixed to eliminate the rigid body displacement of the model. First, elastic stress analysis without geometric nonlinearity was performed to

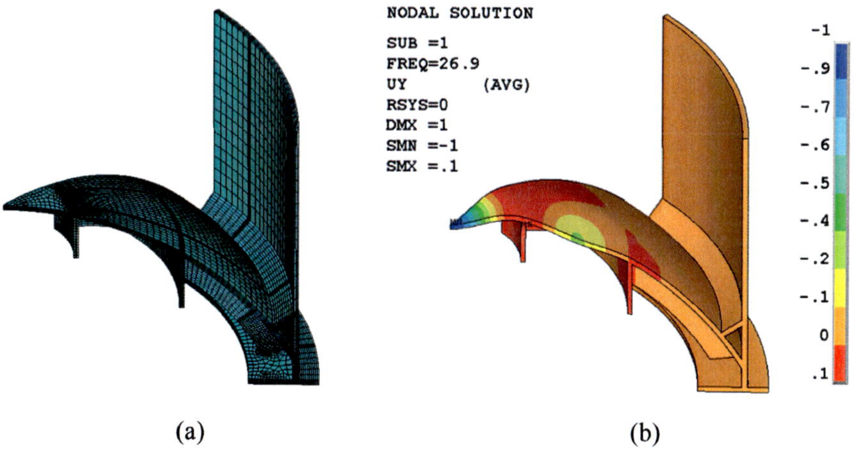

Fig. 2.8 Protection against buckling for spherical head: **a** Mesh model; **b** Buckling mode (from Refs. [7, 10] with permission of ASME)

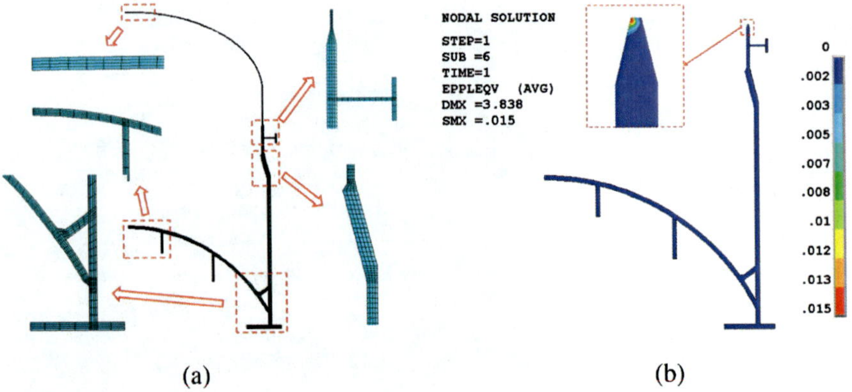

Fig. 2.9 Stress analysis for test vessel: **a** Mesh model; **b** Plot of von Mises plastic strain at design pressure (from Refs. [7, 10] with permission of ASME)

determine the pre-stress in the spherical head. Second, eigenvalue buckling analysis was performed to determine the minimum eigenvalue ($e_{\min}$). Figure 2.8b shows the buckling mode of the model, and $e_{\min}$ was determined as 26.9. Design factor (F_{D}) was 16.13 for the spherical head under external pressure in accordance with Part 5.4 of ASME VIII-2. The allowable buckling pressure (P) of the spherical head was determined by $P = e_{\min} P_{\mathrm{d}} / F_{\mathrm{D}}$, and thus $P = 5.0$ MPa. Since the allowable buckling pressure P of 5.0 MPa was greater than the design pressure P_{d} of 3.0 MPa, the spherical head of the test vessel was acceptable with respect to buckling under the design pressure.

Due to symmetry in geometry and loading, an axisymmetric plane model of the test vessel with the ZJU-FA head was generated. The geometry of the model is given in Fig. 2.5a. The model was meshed using 2D, 8-node structural solid element PLANE82, as shown in Fig. 2.9a. To avoid plastic collapse of the ZJU-FA head and Cylinder A under the design pressure of 3 MPa, the material of the ZJU-FA head and Cylinder A was assumed to have a higher yield strength of 830 MPa. Because the thickness of the remaining components of test vessel was more than four times those of the ZJU-FA head and Cylinder A, the remaining components were in an elastic state when the ZJU-FA head ruptured. Therefore, the material Q345R of the remaining components was assumed to be elastic perfectly plastic with a Young's modulus of 206 GPa, Poisson's ratio of 0.3 and yield strength of 330 MPa per GB/T 150.2 [14]. The design pressure of 3 MPa was applied to the pressure boundaries of the test vessel. Radial displacement constraints were applied to the symmetric axis of the test vessel and an axial displacement constraint was applied to the bottom edge of the base board. A stress analysis for the design pressure was performed and a plot of the von Mises plastic strain is shown in Fig. 2.9b. It can be seen from Fig. 2.9b that the local plastic deformation at the joint of the ellipsoidal head and the transition shell was within a very small range, and zero plastic strain was incurred in the remaining components including transition shell with T type reinforced ring,

Fig. 2.10 Test vessel (from Refs. [7, 10] with permission of ASME)

conical shell, cylindrical shell (50 mm), conical reinforced shell, spherical head or base board.

Therefore, the above bucking evaluation and stress analysis show that the test vessel can be reused for the buckling experiments of large ellipsoidal heads. A picture of the test vessel with ZJU-FA head after fabrication is shown in Fig. 2.10.

2.2.2 Measurement of Initial Shape and Deformation with 3D Laser Scanner

Linear voltage displacement transducers (LVDTs) are usually used to measure the initial shape and deformation of a structure; for example, Miller et al. [15] performed failure tests on two torispherical heads with a diameter of 4877 mm, then carried out shape measurement for the two torispherical heads using a specially designed measuring device with fifteen LVDTs (Fig. 2.11). As the measuring arm was rotated around the heads at three increments, the transducers measured displacement relative to the surface of the measuring arm; thus, any change in the radial dimension could represent a change in shape from previous measurements. Obviously, LVDTs only measure displacement in one direction at one point of a structure, so it is difficult to

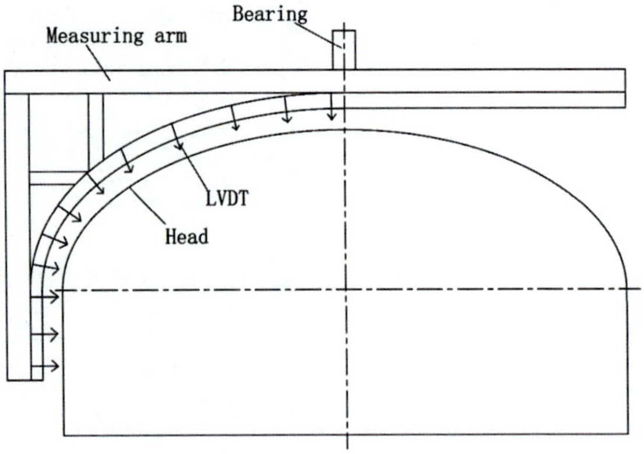

Fig. 2.11 Shape-measuring device with LVDTs for large heads tested by Miller et al. [15] (from Ref. [10])

obtain the overall shape of a structure using LVDTs. Additionally, measuring devices need to be specially designed to mount LVDTs, as shown in Fig. 2.11, further adding to the complexity of the task of measuring the shape.

With the development of optical measuring technology, using a 3D laser scanner has become a priority for shape measurement. Recently, some researchers [16–19] used 3D laser scanner to measure the shapes of shells and determined their shape imperfections. The 3D laser scanner is a noncontact measuring instrument which can easily determine all coordinates of a large structure. To measure the initial shape and deformation of the tested ellipsoidal heads with diameters of 4797–5000 mm, we used a terrestrial 3D sensor with a maximum resolution of 1 mm, maximum accuracy of 0.5 mm, and maximum measurement rate of $\leq$1 million pixels/second. The sensor operates within a maximum range of 187 m, considerably reducing the high risk of sensor damage due to bursting of the tested ellipsoidal heads.

During the test, the 3D laser scanners were supported by steel frames and located slightly above the ellipsoidal heads (Fig. 2.12). The point cloud measurement data shown in Fig. 2.13 represents the output from the 3D laser scanner. The actual shape of the measured tested ellipsoidal head (Fig. 2.14) was modelled from the point cloud data. The outside diameters and heights of the tested ellipsoidal heads given in Table 2.3 were measured directly from the actual structures.

In addition, the thicknesses of the tested ellipsoidal heads were measured with ultrasonic thickness gauges. The average thicknesses of the crowns and knuckles are also given in Table 2.3. The average measured thicknesses of cylinders A and B attached to the tested ellipsoidal heads (Fig. 2.5) were 8.3 mm and 8.6 mm respectively.

Fig. 2.12 Shape measurement for test heads using 3D laser scanner (from Ref. [8] with permission of Elsevier)

Fig. 2.13 Point cloud data from 3D laser scanner (from Ref. [10])

Fig. 2.14 Actual shape of ellipsoidal head using 3D laser scanner (from Ref. [10])

Table 2.3 Geometry measurements of tested large ellipsoidal heads (from Ref. [10])

Head No	Outside diameter (mm)	Outside height (mm)	Thickness (mm)	
			Crown	Knuckle
ZJU-FA1	4811.9	1407.5	5.7	6.1
ZJU-FA2	4817.7	1412.8	5.7	5.9
ZJU-FT1	5013.7	1284.3	5.8	6.1
ZJU-FT2	5011.4	1275.1	5.8	6.0
ZJU-FN1	5013.4	1612.5	5.9	6.2
ZJU-FN2	5008.6	1585.3	6.0	5.9

2.2.3 *Measurement of Large Strain Under Hydraulic Pressure*

When a buckle forms, a large bending deformation occurs in the buckle, resulting in large strain and a significant strain difference between the inside and outside surfaces of the buckle. Strain measurement is a good method for obtaining detailed buckling deformation. A strain gauge is bonded to the measuring object with a dedicated adhesive. The strain on the measuring site is transferred to the strain sensing element via the gauge base, thus changing the electrical resistance of the element. Electrical resistance changes are read out on an all-in-one measuring instrument, i.e. a data logger fully controlled from a computer.

Since the test heads were subjected to hydraulic pressure, adding the difficulty of measuring large strain at the inside surface of heads in this condition. In order to measure large buckle strain, we chose an ultrahigh elongation foil strain gauge (designation: KFEM-5–120-C1) with a uniaxial gauge pattern (Fig. 2.15), which has an ultrahigh elongation of 20–30%. In addition, the lead-wire cable of strain gauge

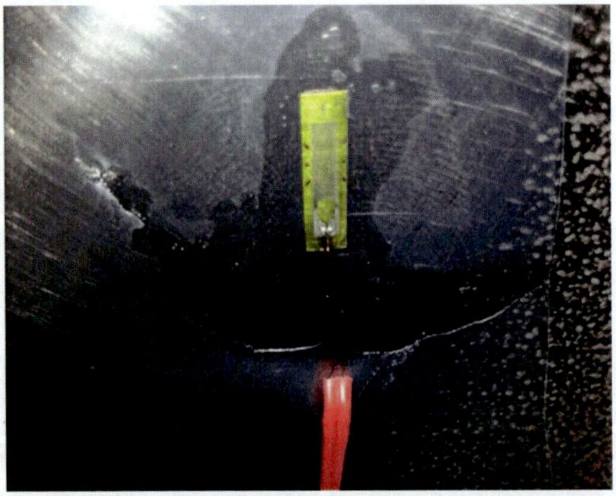

Fig. 2.15 Ultrahigh elongation foil strain gauge (from Ref. [10])

had a length of 5 m, reducing the number of seals in the wire joint. Strain gauges on the inside surfaces were waterproofed with silicone rubber (Fig. 2.16) due to its water resistance and ultrahigh elongation of up to 250%. Wire joints between strain gages and sealed feedthrough assembly were sealed with plastic pipes filled with silicone rubber, as shown in Fig. 2.17. Sealed feedthrough assemblies were specially designed to obtain strain data from inside of the test vessel under hydraulic pressure. As shown in Fig. 2.18, the assembly consists of a swaged stainless steel tube and insulated wires to form a continuous wire feedthrough. The assembly permitted multiple insulated wires to be installed through a single port and pass through a pressure boundary to prohibit the leakage of gas/liquid media.

Because buckles occur at the head knuckle, the strain gauges were attached at the knuckle of the test head along the circumferential direction (Fig. 2.19). At each location, two strain gauges (one in the meridional direction and the other in the

Fig. 2.16 Strain gauges sealed with silicone rubber (from Ref. [10])

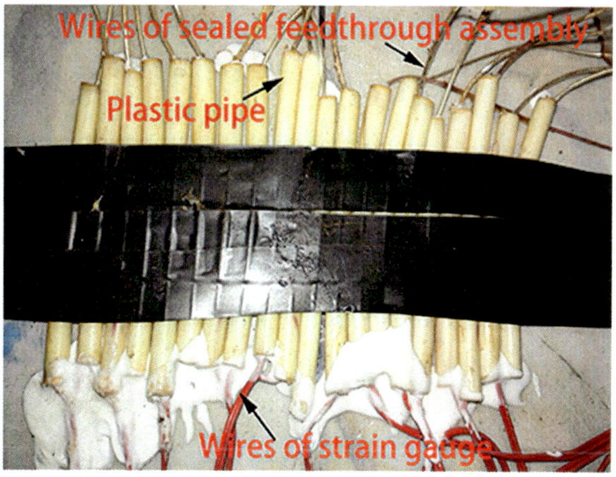

Fig. 2.17 Wire joints sealed with plastic pipes (from Ref. [10])

Fig. 2.18 Sealed feedthrough assembly (from Ref. [10])

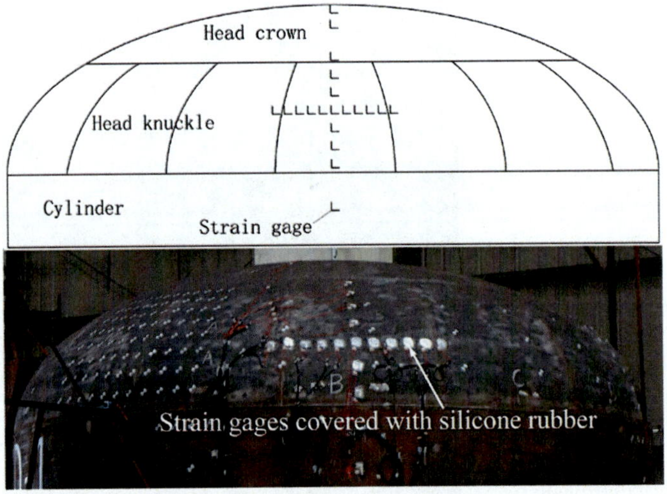

Fig. 2.19 Location of strain gauges (from Ref. [10])

circumferential direction) were attached to the inside surfaces at the same locations corresponding to another two strain gauges outside.

The pressurizing and measuring system is shown schematically in Fig. 2.20. Having assembled and instrumented the test vessel, it was filled with water and pressurized with a pump. The test proceeded as follows:

(1) There was an initial scan of the strain gauges and laser scanners at "zero" pressure.

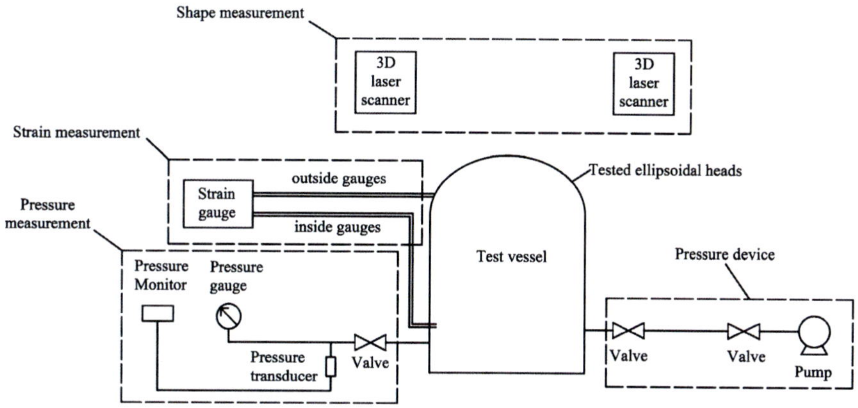

Fig. 2.20 Pressuring and measuring system (from Ref. [7] with permission of ASME)

(2) The pressure slowly increased to a suitable value and held constant, then 3D laser scanners were used to obtain detailed shape measurements of the ellipsoidal head. Because of the high risk of laser sensor damage due to the rupture of the test heads, deformation measurements were not performed with 3D laser sensors above a pressure of 0.95 MPa (ZJU-FT heads) and 1.45 MPa (ZJU-FA heads). Pressure and strain values were measured all the time.

(3) Above 0.95 MPa or 1.45 MPa, the test vessels were pressurized continuously.

2.3 Finite Element Models for Buckling Simulation

2.3.1 Heads with Perfect Shape

Nonlinear finite element method was used to simulate the buckling behavior of the tested ellipsoidal heads. Figure 2.21a1 and b1 show the geometrical structures of the FE models of the ZJU-FA and ZJU-FT heads respectively. The geometrical structure of FE model of the ZJU-FA head includes the ZJU-FA head, Cylinder A, transition shell with T type reinforced ring, conical shell and cylindrical shell. The geometrical structure of ZJU-FN head model includes the ZJU-FN head, Cylinder B and cylindrical shell, similar to that of the ZJU-FT head. The models have perfect shapes. The measured diameters, heights and average thicknesses (Table 2.3) were imported into the models of the tested ellipsoidal heads and cylinders. The other components of the models adopt nominal thicknesses, as shown in Fig. 2.5a.

Element SHELL181 was used to generate the mesh models, as shown in Fig. 2.21a2 and b2. SHELL181 is a 4-node structural shell element for analyzing thin to moderately thick shell structures, and is well suited to linear, large deflection and large strain nonlinear applications. The measured true stress–strain curves (Fig. 2.4) were used as the material models of the tested ellipsoidal heads. Since

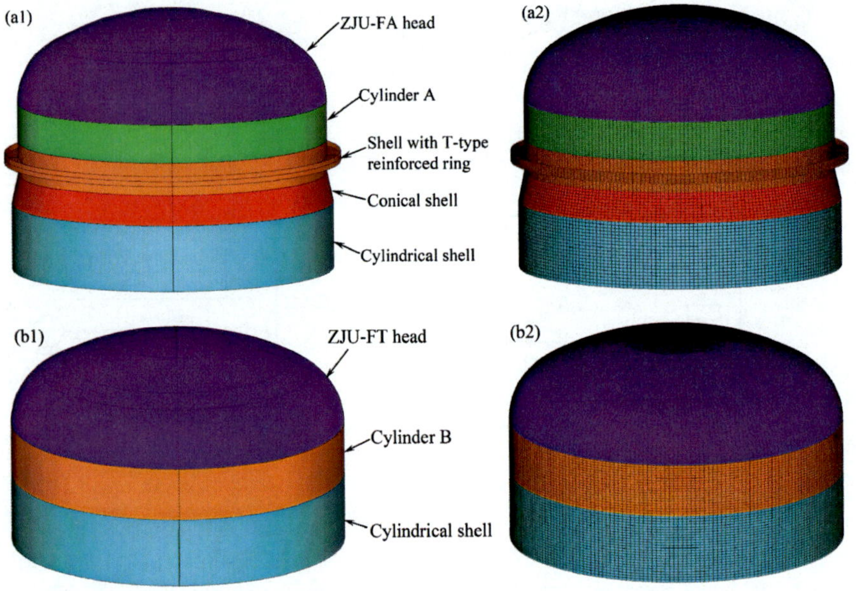

Fig. 2.21 **a1** Geometric model and **a2** mesh model for ZJU-FA head; **b1** Geometric model and **b2** mesh model for ZJU-FT head (from Ref. [9] with permission of Elsevier)

the thicknesses of the other components are much greater than those of the tested ellipsoidal heads, the material Q345R of the other components was assumed to be elastic-perfectly plastic with Young's modulus of 206 GPa, Poisson's ratio of 0.3 and measured yield strength of 330 MPa. The flow theory of plasticity was adopted in the material models. The FE models were loaded by applying internal pressure on their inner surface, and they were fully clamped at the bottom of the cylindrical shells. The cylindrical shells were modelled with such a length that the effect of edge stress was negligible. The FE models were analyzed using the arc-length method [20–22], taking into account material and geometrical nonlinearities.

2.3.2 Heads with Actual Shape

It is known that shape imperfections affect the buckling of structures. Recently, some researchers [17–19] carried out the research on FE models for buckling based on the measured shapes. As mentioned in Sect. 2.2.2, 3D laser scanners were used to measure the initial shapes of the tested ellipsoidal heads with cylinders. The measured initial shapes were introduced into the FE models to consider the effect of the shape imperfections. Figure 2.22 presents the FE models for the heads with actual shape in the case of ZJU-FN head. The other parameters of the FE models with actual shape are the same as those of the models with perfect shape in Sect. 2.3.1.

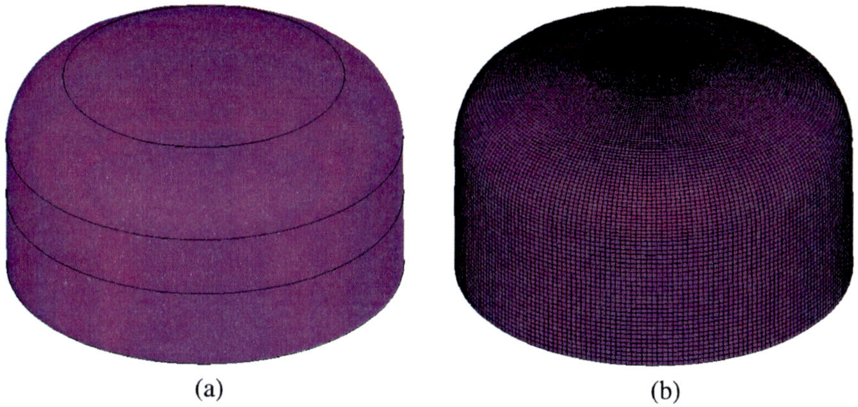

Fig. 2.22 **a** Geometrical model and **b** mesh model for heads with actual shape (from Ref. [10])

2.4 Buckling Behavior

2.4.1 Determination of Local Buckling Pressure

The experimental and FE results show that buckles formed in the meridional direction of the head knuckle, with a very short circumferential wavelength, as shown in Figs. 2.23 and 2.24. It is evident that the ZJU-FA2 and ZJU-FT2 heads experienced local buckling. Some buckles occurred at weld joints, while others (e.g., the buckle at 86° in the ZJU-FA2 head) did not. Only one type of buckling pattern was observed in which the buckles formed outward. However, buckles are not observed in the ZJU-FN head experiments.

The local buckling pressure of an internally pressurized ellipsoidal head is the pressure that causes the onset of the first buckle in the head. In this study, two methods were used to determine the experimental local buckling pressure [4, 23–25]. First, circumferential strain was used if strain gauges were located at the first buckle. Second, 3D laser scanning was used to determine changes in head shape.

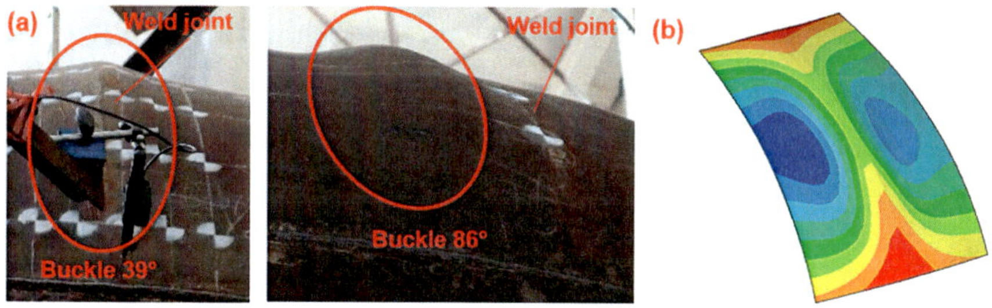

Fig. 2.23 View of typical buckles in ZJU-FA2 head: **a** Experimental; **b** FE results (from Ref. [9] with permission of Elsevier)

Fig. 2.24 View of typical buckles for ZJU-FT2 head: **a** Experimental; **b** FE results (from Ref. [9] with permission of Elsevier)

Figure 2.25 shows the measured circumferential strain-pressure curves for the first buckle in the ZJU-FT2 head. After test pressure of 700 kPa, the circumferential strains of the inner and outer surfaces diverged tremendously, and the inner surface changed from compressive to tensile strain, while the compressive strain of the outer surface became larger; this was because inward circumferential bending occurred due to the formation of the buckle. Therefore, the local buckling pressure of the ZJU-FT2 head was measured as 700 kPa. But for other tested heads, there were no strain gauges locating at the first buckle, so we used shape change from 3D laser scanning to determine the buckling pressure. Figure 2.26 shows the evolution of buckles with test pressure for the ZJU-FA2 head, obtained by 3D laser scanning. It can be seen in Fig. 2.26 that at test pressure of 1.15 MPa, there was no buckling, while a buckle was

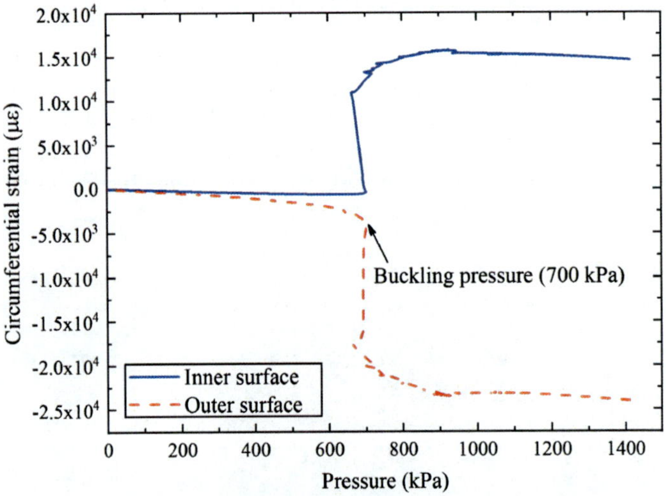

Fig. 2.25 Experimental results of circumferential strain–pressure curves for ZJU-FT2 head (from Ref. [9] with permission of Elsevier)

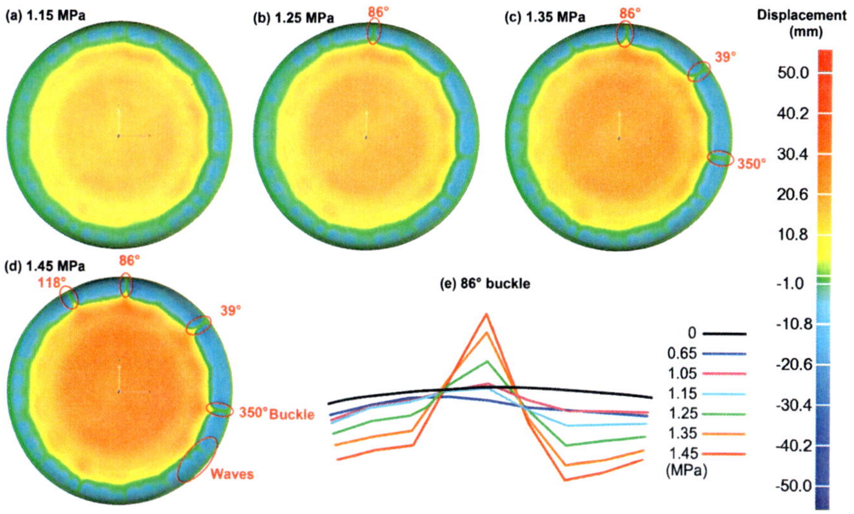

Fig. 2.26 Experimental results of buckling evolution of ZJU-FA2 head (from Ref. [9] with permission of Elsevier)

observed at the 86° location when the test pressure was 1.25 MPa (Fig. 2.26e). As the head was further pressurized, other buckles occurred around the circumference, as shown in Fig. 2.26c and d. Therefore, the buckling pressure of the ZJU-FA2 head is estimated as 1.15–1.25 MPa. As such, we adopted the lower value of 1.15 MPa as the buckling pressure of the ZJU-FA2 head. Similarly, the buckling pressures of ZJU-FA1 and ZJU-FT1 heads were 1.15 MPa and 0.70 MPa respectively.

It is easier to determine buckling pressure for the FE models. Figures 2.27 and 2.28 show the FE results of the buckling pressure of ZJU-FA1 head with perfect and actual shapes. For the ZJU-FA1 head with perfect shape, the two points of A and B were selected on a buckle along the circumferential direction, as shown in Fig. 2.27a, and the pressure-radial displacement curves of each point are shown in Fig. 2.27b. It

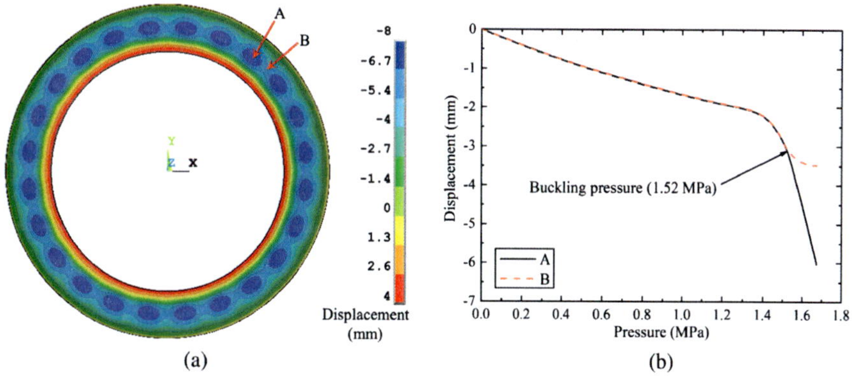

Fig. 2.27 FE results of buckling pressure of ZJU-FA1 head with perfect shape: **a** Buckle pattern; **b** Displacement–pressure curves (from Refs. [7, 10] with permission of ASME)

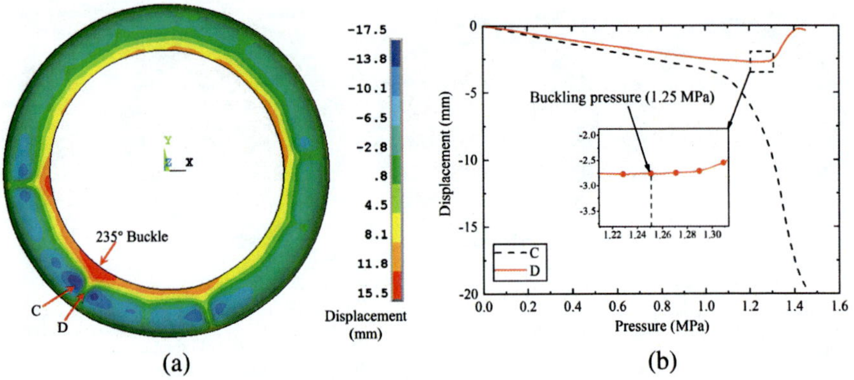

Fig. 2.28 FE results of buckling pressure of ZJU-FA1 head with actual shape: **a** Buckle pattern; **b** Displacement–pressure curves (from Refs. [7, 10] with permission of ASME)

can be seen in Fig. 2.27b that the radial displacements of A and B are almost identical at pressures ranging from 0 to 1.52 MPa. After 1.52 MPa, the displacement value of B is different from that of A, indicating the occurrence of buckling deformation. Therefore, the buckling pressure of head ZJU-FA1 with perfect shape is predicted as 1.52 MPa. For the ZJU-FA1 head with actual shape, the two points of C and D were selected on a buckle along the circumferential direction, as shown in Fig. 2.28a, and the pressure-radial displacement curves of each point are shown in Fig. 2.28b. We can see from Fig. 2.28b that these curves diverge from the beginning of the pressure, and the displacements of C and D become nonlinear after pressure of about 1.0 MPa due to plastic deformation. This is a result of bending deformation due to shape imperfection included in the head model with actual shape. As the pressure exceeds 1.25 MPa, the displacement direction of D changes and becomes different from that of C, which is accepted as the onset of buckling. Therefore, the buckling pressure of the ZJU-FA1 head with actual shape was predicted as 1.25 MPa. Similarly, the predicted buckling pressures of ZJU-FA2, ZJU-FT1 and ZJU-FT2 heads were determined and given in Table 2.4.

Table 2.4 Comparison between predicted buckling pressures and experimental results (from Ref. [10])

Head no	Experimental buckling pressure, P_{exp} (MPa)	Head model with perfect shape		Head model with actual shape	
		Predicted buckling pressure, P_b (MPa)	$\frac{P_b}{P_{exp}}$	Predicted buckling pressure, P_b (MPa)	$\frac{P_b}{P_{exp}}$
ZJU-FT1	0.70	1.13	1.61	0.90	1.28
ZJU-FT2	0.70	1.01	1.44	0.92	1.31
ZJU-FA1	1.15	1.52	1.32	1.25	1.09
ZJU-FA2	1.15	1.72	1.50	1.25	1.09

Table 2.4 shows a comparison of the experimental and predicted buckling pressures for head models with perfect and actual shapes. Compared with head models with perfect shape, the predictions concerning actual shape are in better agreement with the experimental results. This means that initial geometrical imperfection has a significant effect on buckling pressure. The differences between the experimental and FE results may be attributed to the assumption of average shell thickness, homogeneity of material properties used in FE models, the error introduced by obtaining of the measured initial model for FE analysis, etc. In particular, the non-uniformity in thickness and material properties was caused by the forming and welding process, significantly reducing buckling pressure.

2.4.2 *Characteristics of Buckling*

2.4.2.1 Locality

As presented in Sect. 2.4.1, buckling occurs at local positions in the knuckle of the ellipsoidal head. Figure 2.29 shows the distribution of measured circumferential strains along the circumference of the ZJU-FT2 head knuckle after the occurrence of buckle at test pressure reducing to about 670 kPa. It can be seen that the differences between the circumferential strains of the inner and outer surface at the locations of Gauges 19 and 20 (corresponding to the buckle at 300°) are much greater than those of the other gauges, because the formation of the buckle at 300° leads to local bending deformation. This indicates that this kind of buckling has the characteristic of "locality". As mentioned in Sect. 1.2, only circumferential stress in the head

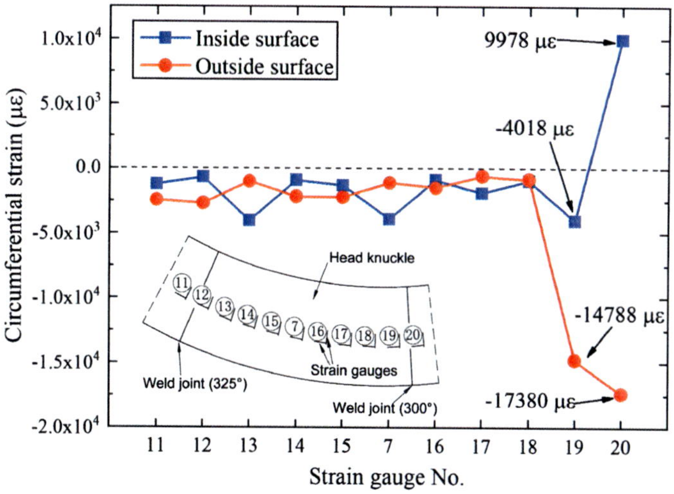

Fig. 2.29 Distribution of measured circumferential strains along circumference of head knuckle for ZJU-FT2 head (from Ref. [9] with permission of Elsevier)

knuckle is negative; that is, head knuckle is in compressive state. This is why buckling is produced in the knuckle for internally pressurized ellipsoidal heads. The local distribution of compressive stresses leads to the buckling characteristic of "locality".

2.4.2.2 Progressivity

As shown in Fig. 2.26, a buckle first occurred at a local position on the head knuckle. As the head was further pressurized, other buckles and small waves progressively formed around the circumference, as shown in Fig. 2.26c and d. After the test, more buckles were observed. The head experienced non-axisymmetric buckling. Figure 2.30 shows the FE results for the evolution of buckles in the ZJU-FA2 head with actual shape, corresponding to the displacement distribution results along a circumference in the knuckle (Fig. 2.31). Similar to the experimental results, some buckles first occurred at local positions in the knuckle, while more formed progressively as the pressure increased. Also, the locations of many buckles shown in Fig. 2.30e are almost same as those observed in the test of the ZJU-FA2 head. This indicates that such buckling has the characteristic of "progressivity".

In the case of the ZJU-FT2 head, the evolution of the buckles with increasing pressure is shown in Fig. 2.32. At 0.72 MPa, two buckles formed at 300° and 351°, as shown in Fig. 2.32b. As the head was further loaded, more buckles progressively formed around the circumference of the knuckle, which is also non-axisymmetric buckling. This indicates that the buckling of the ZJU-FT2 head also showed characteristic of "progressivity". At 0.95 MPa, ten buckles formed, as shown in Fig. 2.32f. More buckles were observed after the test of the ZJU-FT2 head. The buckles in the

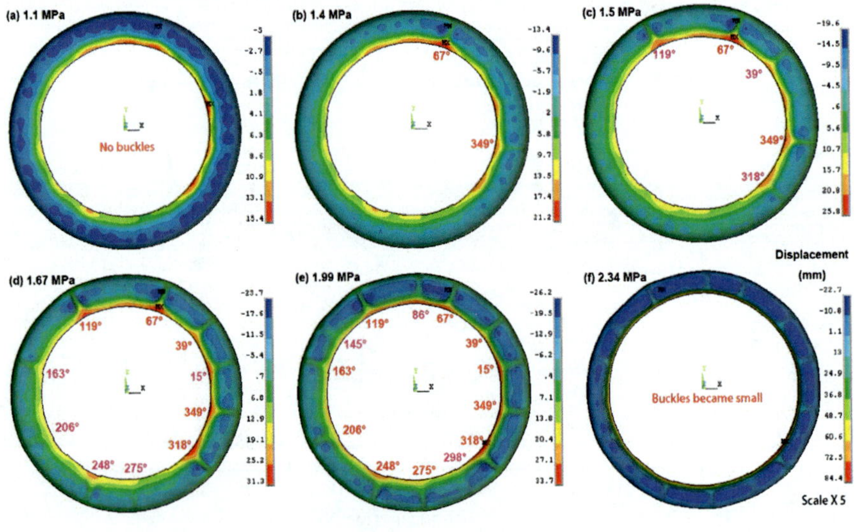

Fig. 2.30 FE simulation of buckling evolution of ZJU-FA2 head (from Ref. [9] with permission of Elsevier)

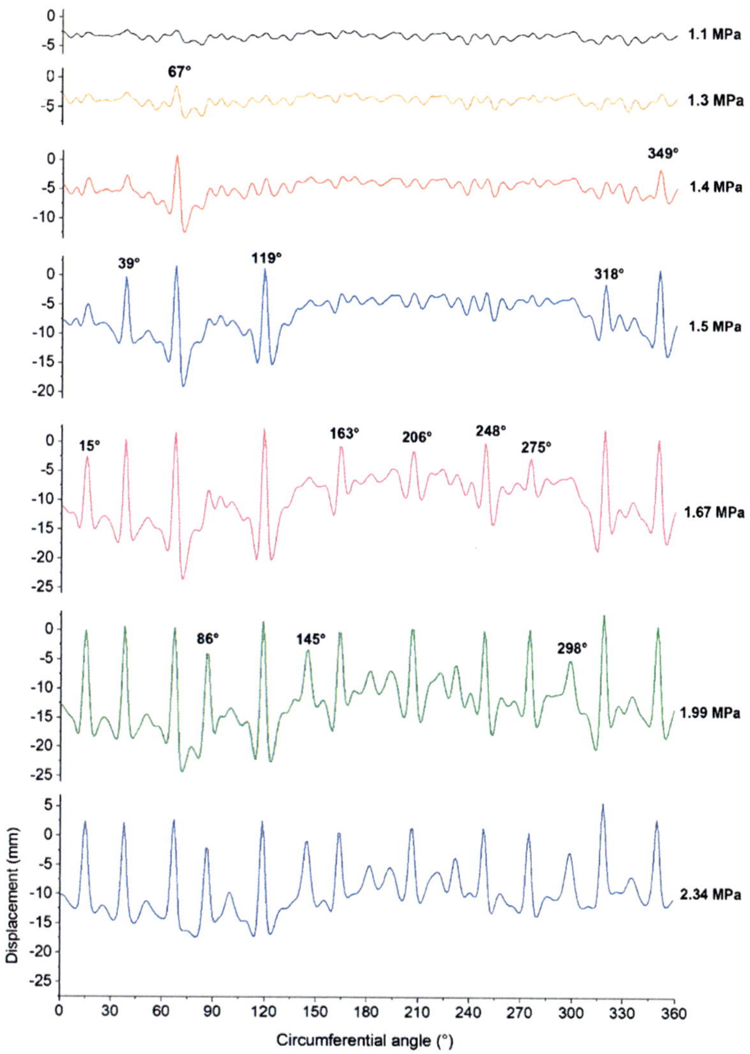

Fig. 2.31 FE simulation on displacement distribution along circumference in knuckle of ZJU-FA2 head (from Ref. [9] with permission of Elsevier)

two ZJU-FT heads were all located at the weld joints of the knuckle. The locations of many buckles obtained by the FE method were almost the same as those observed in the experiments of the two ZJU-FT heads; see Figs. 2.32f and 2.34a.

Figure 2.33a shows the FE results of buckle changes for the ZJU-FT2, ZJU-FA2 and ZJU-FN1 heads with actual shape. Buckle height is defined as the difference in displacement values between a peak and a trough of a buckle. Differences exist in the growth of buckles among the ZJU-FT2, ZJU-FA2 and ZJU-FN1 heads. For the ZJU-FT2 head, the buckles grow very rapidly at first before the growth becomes slow. For the ZJU-FA2 head, the growth of the buckles is initially slow, while the initial growth

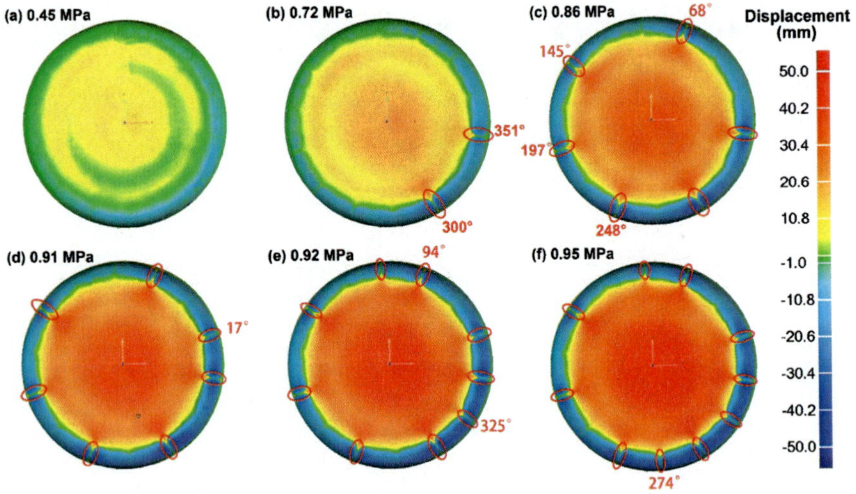

Fig. 2.32 Experimental results of buckle evolution of ZJU-FT2 head (from Ref. [9] with permission of Elsevier)

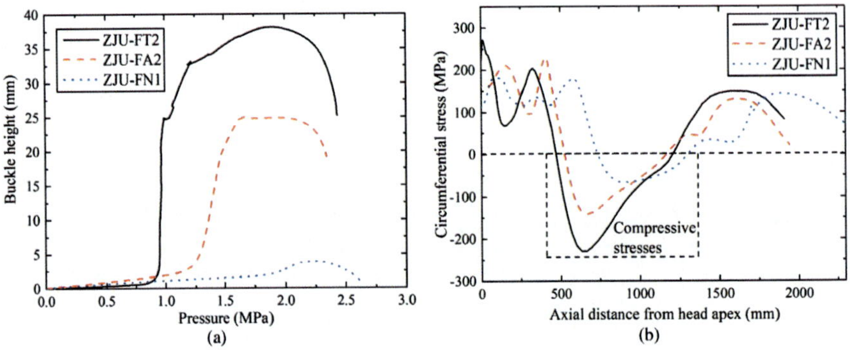

Fig. 2.33 FE results for ZJU-FT2, ZJU-FA2 and ZJU-FN1 heads with actual shape: **a** Curve of buckle height versus pressure; **b** Circumferential stresses in middle surface (from Ref. [9] with permission of Elsevier)

of the buckles is much slower for the ZJU-FN1 head, which means that the growth of a buckle also has a characteristic of "progressivity" for some heads. In addition, the FE results show that a buckle in the ZJU-FN1 head has a maximum height of 4 mm and nearly disappears at relatively higher pressure. Therefore, although such small buckles occurred in the buckling experiment for the ZJU-FN1 head, it is very difficult to measure them with 3D laser scanners.

The initial growth of most buckles in the ZJU-FT2 head was sudden, while those in the ZJU-FA2 head grew gradually. Furthermore, no buckle was observed in the test of the ZJU-FN1 head with a relatively small radius-to-height ratio. In addition, for the ZJU-FA2 head, there was no sudden pressure drop as with the ZJU-FT2 head. This can be directly explained by the comparison of the circumferential stresses in the middle surfaces of ZJU-FT2, ZJU-FA2 and ZJU-FN1 heads. When the three heads were subjected to the same pressure, the range and magnitude of circumferential stresses in the ZJU-FN1 head was much smaller than those in the ZJU-FA2 head and the ZJU-FT2 head (the ZJU-FA2 head was smaller than the ZJU-FT2 head), as shown in Fig. 2.33b. Smaller circumferential stresses can lead to more stable buckling behavior, even to the occurrence of no buckling.

2.4.2.3 Self-limitation

As shown in Fig. 2.33a, the buckle heights of ZJU-FA2 and ZJU-FT2 heads reduced, and the buckles of the ZJU-FN1 head nearly disappeared. Figures 2.30 and 2.31 show the buckling evolution of the ZJU-FA2 head, and a comparison of the results corresponding to 1.9 and 2.34 MPa reveals that the buckles became smaller. In addition, the FE simulation of the buckling evolution of the ZJU-FT2 head is shown in Fig. 2.34. The buckles of the ZJU-FT2 head also became smaller when the pressure increased from 1.94 to 2.43 MPa. For ellipsoidal heads subjected to internal pressure, buckles become smaller or even disappear, indicating that this buckling has a characteristic of self-limitation.

Figure 2.35 shows the distribution of circumferential stresses in the middle surface along a meridian under different pressures in the case of the ZJU-FT2 head. It should be noted that the curves shown in Figs. 2.35 and 2.36 are not smooth because the shape imperfection included in the FE models led to the fluctuation in the circumferential stress distribution. As shown in Fig. 2.35, the compressive stress in the knuckle became smaller when the pressure exceeded a certain value. Furthermore, the stress states of some positions changed from compressive to tensile as pressure

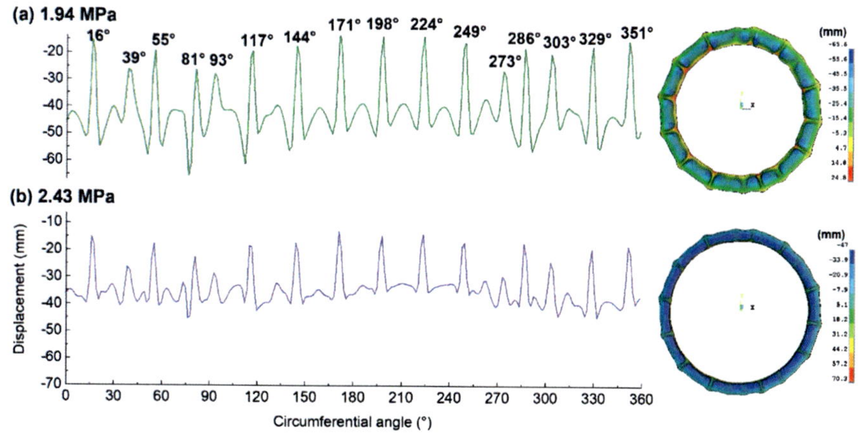

Fig. 2.34 FE simulation of buckling evolution for ZJU-FT2 head (from Ref. [9] with permission of Elsevier)

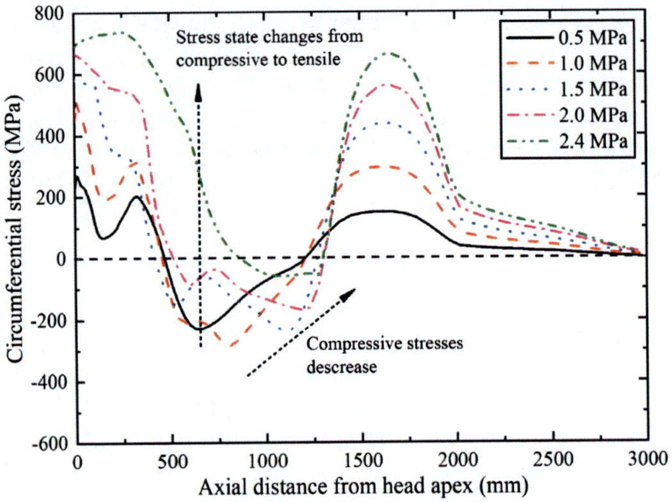

Fig. 2.35 Distribution of circumferential stresses along meridian in ZJU-FT2 head (from Ref. [9] with permission of Elsevier)

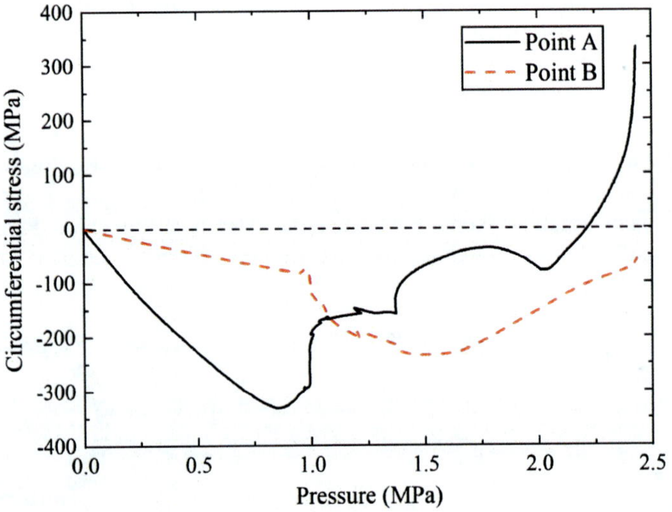

Fig. 2.36 Circumferential stress-pressure curves of points A and B at ZJU-FT2 head knuckle (from Ref. [9] with permission of Elsevier)

increases; for example, the point A shown in Fig. 2.36. It is clear from Fig. 2.36 that the compressive stresses in the knuckle decreased with the increasing pressure. Therefore, when the pressure was relatively higher, buckling was self-limited in the internally pressurized ellipsoidal heads.

The change in the shape of the ZJU-FT2 head is illustrated in Fig. 2.37. It can be seen from Fig. 2.37a that head crown and cylinder deform outside while head

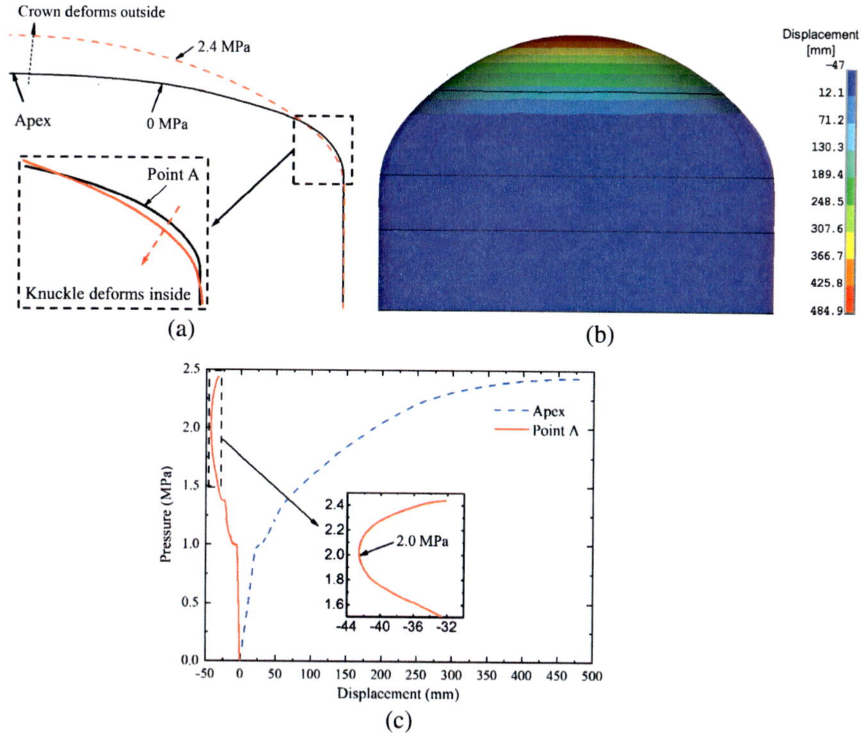

Fig. 2.37 Deformation of ZJU-FT2 head: **a** Shape change along meridian; **b** Contour of displacement at 2.4 MPa; **c** Pressure–displacement curves (from Ref. [9] with permission of Elsevier)

knuckle deforms inside, leading to the occurrence of compressive circumferential stresses in the knuckle. As shown in Fig. 2.37b, the ellipsoidal head tends to be a hemisphere as a result of internal pressurization, indicating that this head experienced geometric strengthening, as discussed in Sects. 3.2.3 and 3.3.2. In other words, the displacement of the apex was far greater than those of the points in the knuckle, as shown in Fig. 2.37a and c; moreover, Point A of the knuckle first deformed inside, but then began to deform outside when the pressure reached 2.0 MPa; see Fig. 2.37c. In brief, this "hemisphere" deformation can cause a reduction in circumferential stresses; see Figs. 2.35 and 2.36.

Based on the above experimental and FE results, we can summarize the buckling characteristics of internally pressurized ellipsoidal heads. Buckles first form at local positions in the head knuckle. As the pressure increases, more buckles form around the circumference of the knuckle, which is a progressive process. Moreover, the growth of buckles is progressive but not sudden for some heads. In addition, buckles become smaller and even disappear at relatively higher pressure. In summary, the buckling of ellipsoidal heads under internal pressure has three characteristics: locality, progressivity and self-limitation.

2.4.3 *Influencing Factors of Buckling*

In this section, in order to investigate the effects of material properties, diameter-to-thickness ratio (D_i/t), rigidity due to attached component, radius-to-height ratio ($D_i/2h_i$) and shape imperfection on the buckling behavior of internally pressurized ellipsoidal heads, FE models of ellipsoidal heads with cylinders were generated for simplicity, similar to the FE model shown in Fig. 2.6b. The bottoms of the cylinders of these FE models were assumed to be fully clamped. In addition, it was assumed that the ellipsoidal heads and cylinders have the same diameter of 5000 mm and the same thickness. The length (h_c) of the cylinders was assumed to be 500 mm. Buckling analyses on these FE models were performed using the arc-length method, taking material and geometric nonlinearities into account.

2.4.3.1 Yield Strength and Strain Hardening

For the case of perfect head models with $D_i/2h_i = 2.0$ and $D_i/t = 1000$, three kinds of materials (Fig. 2.38) were used to study the effects of yield strength and strain hardening on buckling behavior of ellipsoidal heads. These materials had the same Young's modulus of 200 GPa and Poisson's ratio of 0.3. Materials A and B had the same yield strength of 200 MPa. The material A was assumed to be elastic-perfectly plastic while the material B included strain hardening. The material C has a similar strain hardening behavior; however, it has a greater yield strength of 600 MPa compared to that of the material B.

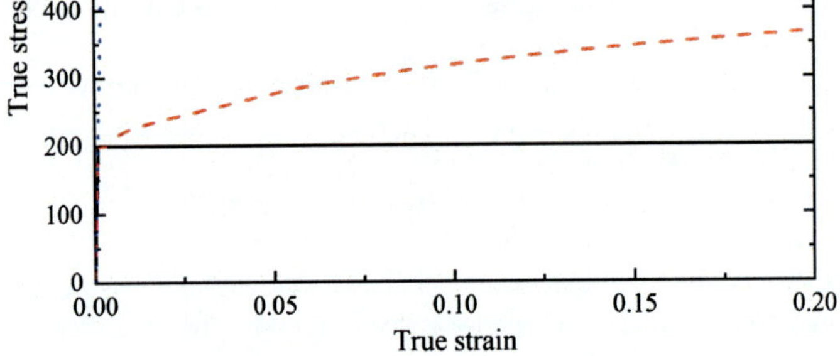

Fig. 2.38 Three kinds of materials for FE models (from Ref. [9] with permission of Elsevier)

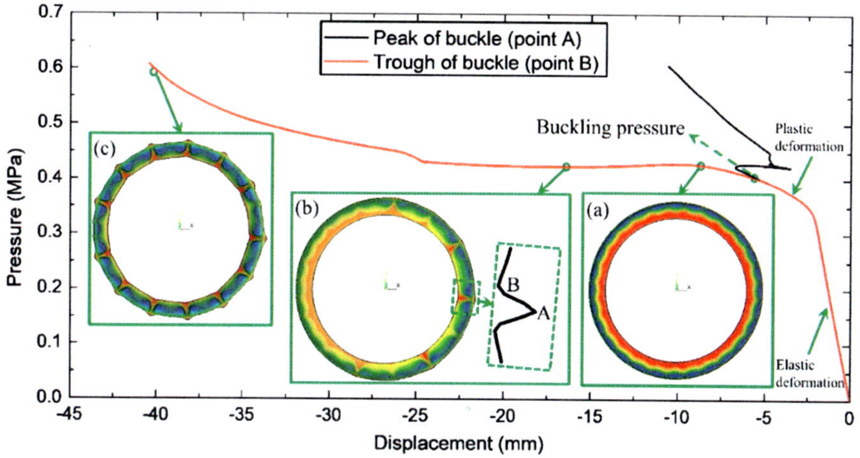

Fig. 2.39 FE results on the buckling behavior of ellipsoidal head (from Ref. [5] with permission of Elsevier)

Figure 2.39 illustrates the FE result on the buckling behavior of the ellipsoidal head with the material model A. We can obtain the pressure–displacement curves of the points A and B corresponding to the peak and trough of a buckle, as shown in Fig. 2.39b. The evolution of the buckling pattern is also presented in Fig. 2.39a–c. It is shown that the displacements are negative in value, that is, the head knuckle deforms inward, causing circumferential compression of the head knuckle. It is this circumferential compression that can cause the occurrence of the buckles along the circumference of the head knuckle. Since the deformation of the head is axisymmetric at lower pressure, the displacements are identical for the points A and B. However, when the pressure exceeds 0.40 MPa, the displacements of the points A and B start to develop diversely, and a type of axisymmetric buckling pattern appears, as shown in Fig. 2.39a, which is bifurcation buckling. As pressure increases, the point A deforms from inward to outward at a pressure of about 0.42 MPa, and then there is a slight drop in pressure, while the displacement of the point B has a considerably increase in value. It is found that there is a transition in the buckling pattern: some bifurcation buckles grow to be remarkable ones shown in Fig. 2.39b. The ellipsoidal head experiences post-buckling behavior. At higher pressure more remarkable buckles occur around the circumference of the head knuckle, as shown in Fig. 2.39c.

As shown in Fig. 2.39, the pressure–displacement curves are linear at lower pressure because elastic deformation appears in the head knuckle. As pressure exceeds 0.33 MPa, the pressure–displacement curves become nonlinear, indicating that plastic deformation occurs. Subsequently, the ellipsoidal head begins to buckle at the pressure of 0.40 MPa. It is obvious that this buckling occurs in the plastic region of material. Furthermore, the spread of plastic strain in this ellipsoidal head is depicted in Fig. 2.40 for several levels of pressure. It is seen that the plastic strains begin on the inner surface of the wall of the knuckle. As pressure increases, the plastic

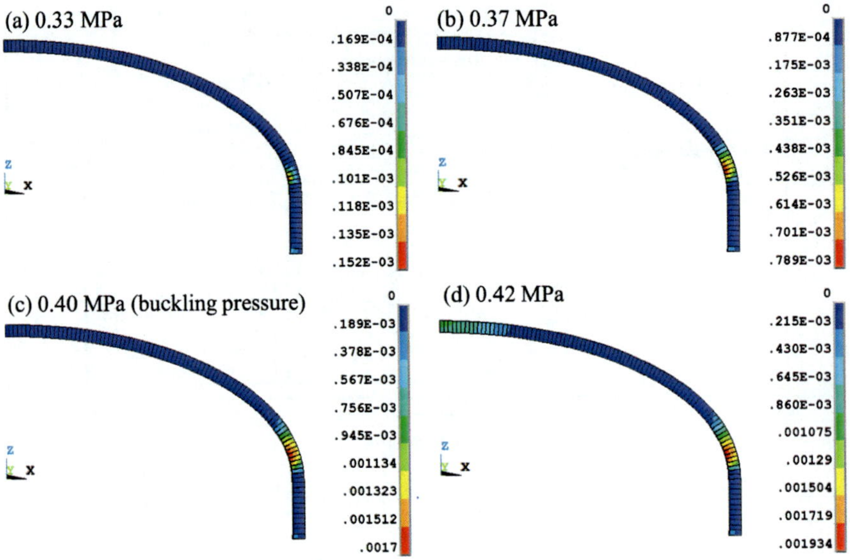

Fig. 2.40 Spread of plastic strain in ellipsoidal head at different pressures (from Ref. [5] with permission of Elsevier)

strains develop across the wall, as shown in Fig. 2.40b. Obviously, the yielding occurs before the buckling pressure (0.40 MPa). With further increase of pressure, the yielding also occurs at the head crown; see Fig. 2.40d. Therefore, plastic bifurcation buckling behavior appears in such ellipsoidal head, as also mentioned in Refs. [2, 26–28].

Figure 2.41 shows the buckle height-pressure curves of the head models with the three kinds of materials. It can be seen that the buckling pressure of the head model with the material C was greater than those of the head models with the materials A and B. Buckling pressure became greater as the yield strength increased because buckling occurred in the plastic range of the material models. In addition, since the material A did not include strain hardening behavior, the "hemisphere" deformation (Sect. 2.4.2.3) of this head model was not significant. Therefore, the reduction in the buckle height of the head model with the material A was very small compared to those of the head models with the materials B and C.

To investigate the effect of material strain hardening on buckling pressure, the FE models for ellipsoidal heads with perfectly plastic material A and material B including strain hardening were generated for different ratios of diameter-to-thickness and radius-to-height, and the FE results of buckling pressures are shown in Table 2.5. For the case of $D_i/2h_i = 2.6$ and $D_i/t = 500$, buckling is predicted to occur for perfect plastic material, but no buckle when strain hardening is considered, which is similar to the case of $D_i/2h_i = 2.0$ and $D_i/t = 750$. For $D_i/2h_i = 2.6$, the buckling pressure obtained by considering strain hardening is higher than that based on perfect plastic material when $D_i/t = 750$. These show that strain hardening can

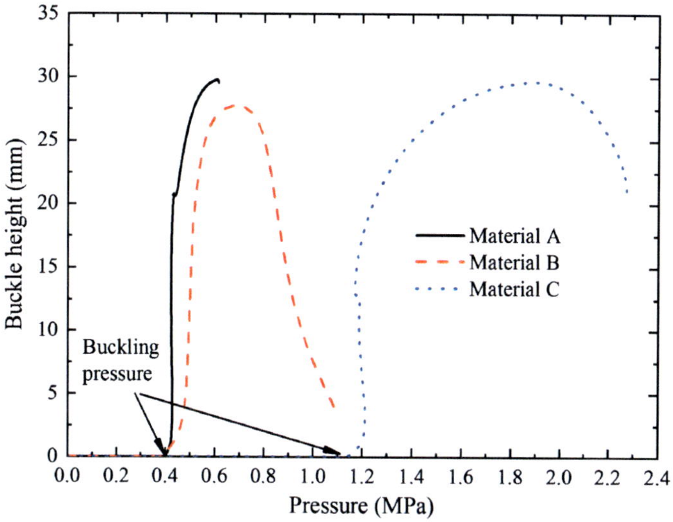

Fig. 2.41 Buckle height-pressure curves of head models with three kinds of materials (from Ref. [9] with permission of Elsevier)

Table 2.5 Comparison of buckling pressures of ellipsoidal heads with perfectly plastic material A and material B including strain hardening (from Ref. [5] with permission of Elsevier)

D_i/t	Buckling pressure, P_b (MPa)			
	$D_i/2h_i = 2.0$		$D_i/2h_i = 2.6$	
	Perfectly plastic material A	Material B including strain hardening	Perfectly plastic material A	Material B including strain hardening
500	NB	NB	0.63	NB
750	0.62	NB	0.33	0.36
1000	0.40	0.40	0.22	0.23

NB = No buckling

cause an increase in resistance to buckling when D_i/t is smaller. However, strain hardening has little or no effect on buckling pressure as D_i/t increases, for example, for $D_i/t = 1000$.

2.4.3.2 Diameter-to-Thickness Ratio

FE head models with $D_i/2h_i = 2.0$ and the material B were used to investigate the effect of diameter-to-thickness ratio (D_i/t) on buckling behavior of ellipsoidal heads. Figure 2.42a shows the buckling behavior of the FE models with different diameter-to-thickness ratios. As D_i/t decreased, buckling pressure increased, and

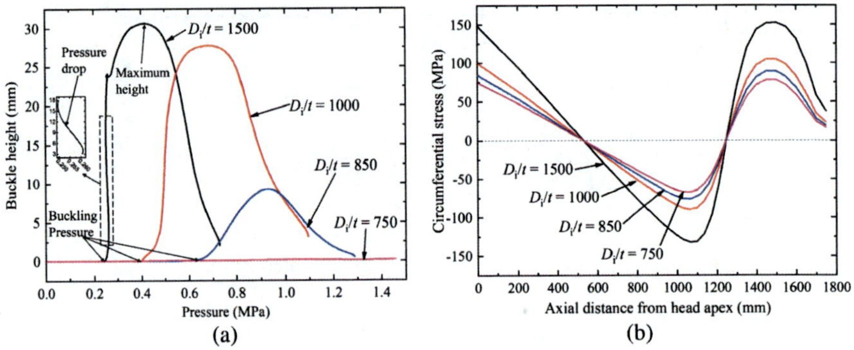

Fig. 2.42 FE results for perfect ellipsoidal heads with different diameter-to-thickness ratios: **a** Curves of buckle height versus pressure; **b** Distribution of circumferential stress (from Ref. [9] with permission of Elsevier)

there was a reduction in the maximum height of the buckles. There were no buckles when $D_i/t = 750$. The deformation of the head models was nearly axisymmetric at lower pressure as a result of perfect shape, hence the buckle height was zero, indicating that buckling did not occur. A buckle formed when pressure exceeded a certain value (buckling pressure), as shown in Fig. 2.42a. Buckles grew slowly at first, but with increasing pressure, the growth of buckles was different depending on D_i/t. In the case of $D_i/t = 1500$, the buckles suddenly became large and a pressure drop occurred. For $D_i/t = 1000$, there was no pressure drop and the buckles grew gradually, while the buckles grew more gradually for $D_i/t = 850$. These changes in buckling behavior are associated with the distribution and magnitude of compressive stresses. As shown in Fig. 2.42b, smaller D_i/t can lead to the smaller value of circumferential stresses in the middle surface under the same pressure level, causing more stable buckling behavior, even to the occurrence of no buckling.

Figure 2.43 shows the buckling evolution of 2:1 ellipsoidal heads with different diameter-to-thickness ratios. For $D_i/t = 850$, the small initial waves gradually became large but there was no transition in the buckling pattern; see Fig. 2.43b1 and b2. In the case of $D_i/t = 1000$, a buckle initially grew slowly, corresponding to the buckling pattern of similar small "waves" (Fig. 2.43a1), but as the pressure increased, a small wave grew into a large buckle (Fig. 2.43a2), and the pattern of this large buckle is close to that observed in our experiments. However, this buckling pattern transition is not observed in our experimental results because the occurrence of buckles is associated with shape imperfection resulting from the manufacturing process.

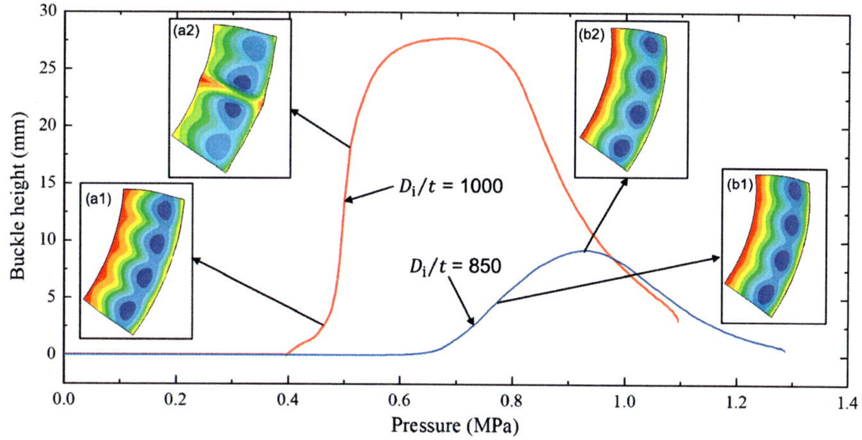

Fig. 2.43 Buckling evolution of perfect head models with different diameter-to-thickness ratios (from Ref. [9] with permission of Elsevier)

2.4.3.3 Rigidity Due to Attached Component

Ellipsoidal heads are attached to cylinders in many engineering applications, so a cylinder is considered in the FE models described above for the buckling simulation of ellipsoidal heads. However, there are some application cases of an ellipsoidal head with a bolting flange. To predict the buckling pressure of an ellipsoidal head with a bolting flange, a cylinder is not considered in the FE models, and the edge of the ellipsoidal head is fixed instead with a bolting flange instead. Table 2.6 lists comparisons of the buckling pressures of ellipsoidal heads with cylinders and without cylinders. In the case of $D_i/2h_i = 3.0$, the buckling pressures of ellipsoidal heads without cylinders are greater than those of ellipsoidal heads with cylinders. The reason is that compared with an ellipsoidal head with a cylinder, an ellipsoidal head without a cylinder is more reinforced by the "fixed" constraint, causing an increase in buckling pressure. It can also be seen that the effect of this reinforcement on buckling pressure decreases with the increasing D_i/t. In addition, in the case of $D_i/2h_i = 2.0$,

Table 2.6 Comparison of buckling pressures of ellipsoidal heads with and without cylinders (from Ref. [5] with permission of Elsevier)

D_i/t	Buckling pressure, P_b (MPa)			
	$D_i/2h_i = 2.0$		$D_i/2h_i = 3.0$	
	With cylinder	Without cylinder	With cylinder	Without cylinder
500	NB	NB	0.46	0.52
1000	0.40	0.41	0.17	0.19
2000	0.18	0.18	0.071	0.075

NB=No Buckling

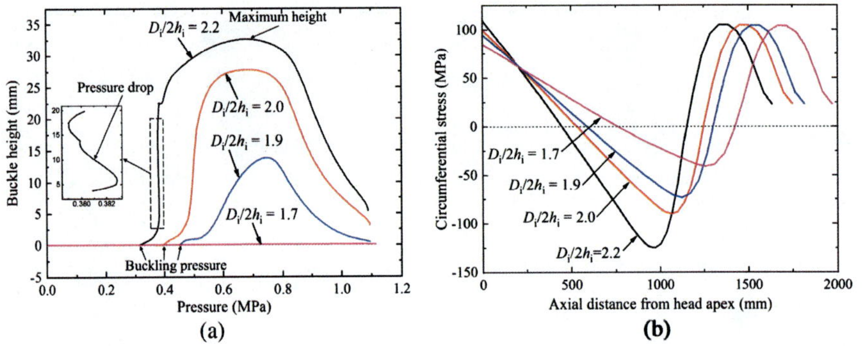

Fig. 2.44 FE results for perfect ellipsoidal heads with different radius-to-height ratios: **a** Curves of buckle height versus pressure; **b** Distribution of circumferential stress (from Ref. [9] with permission of Elsevier)

this effect also decreases as $D_i/2h_i$ decreases. Since the extent of a meridian of an ellipsoidal head with $D_i/2h_i = 2.0$ is greater than that of an ellipsoidal head with $D_i/2h_i = 3.0$, this reinforcement due to the "fixed" constraint has a smaller effect on the circumferential compressive stress associated with the buckling of the ellipsoidal head, leading to a reduction in the difference between the buckling pressures of ellipsoidal heads with and without cylinders.

2.4.3.4 Radius-to-Height Ratio

FE head models with $D_i/t = 1000$ and the material B were used to investigate the effect of the radius-to-height ratio ($D_i/2h_i$) on the buckling behavior of ellipsoidal heads. Similar trends to the effect of the diameter-to-thickness ratio on buckling behavior were obtained for different $D_i/2h_i$, as shown in Fig. 2.44a. As $D_i/2h_i$ decreased, the buckling pressure increased, the maximum height of the buckles reduced, the pressure drop disappeared and the growth of the buckles changed from sudden to gradual. No buckles occurred when $D_i/2h_i = 1.7$. In brief, smaller $D_i/2h_i$ can lead to the smaller value of circumferential stresses in the middle surface under the same pressure (Fig. 2.44b), causing more stable buckling behavior, even to the occurrence of no buckle.

2.4.3.5 Initial Shape Imperfection

The knuckle of a large-scale ellipsoidal head is usually welded by several plates, for example, the tested ellipsoidal heads (Sect. 2.2.1.1). As presented in Fig. 5.13 of Sect. 5.2.2, a bulging of weld, arises along the meridian of the knuckle. An imperfect FE model including a bulging of weld was generated to investigate the effect of this initial shape imperfection on the buckling behavior of an internally pressurized

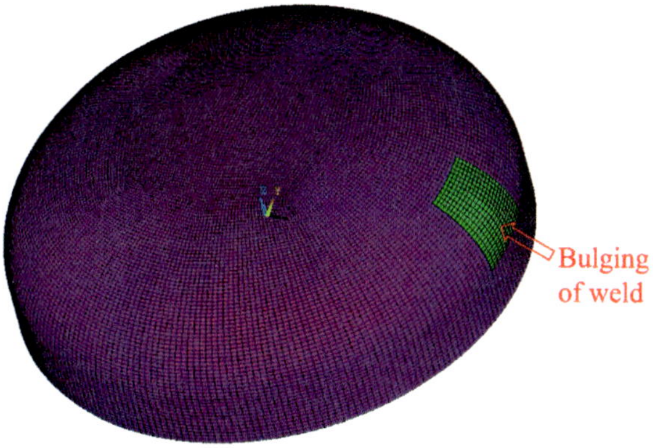

Fig. 2.45 Imperfect FE model including bulging of weld (from Ref. [9] with permission of Elsevier)

ellipsoidal head. Formula (5.1) was proposed in Sect. 5.2.2 to characterize the bulging of welds in head knuckles and it was concluded that the bulging of weld has a considerable effect on the buckling pressure of ellipsoidal heads subjected to internal pressure. The bulging of weld was generated by the following parameters: $\delta_p = 10$ mm, $\varphi_1 = -9°$, $\varphi_2 = 0°$, $\varphi_3 = 9°$, $\theta_1 = 0°$, $\theta_2 = 31.5°$ and $\theta_3 = 63.1°$. This bulging of weld was then introduced into the perfect FE models of ellipsoidal heads with $D_i/2h_i = 2.0$ and $D_i/t = 1000$, as shown in Fig. 2.45. The material B is also used for this imperfect FE model.

Figure 2.46 shows a comparison of the buckling behaviors of the perfect and imperfect head models. It is clear that the buckling pressure of the imperfect model is about 0.33 MPa, lower than that of the prefect model (0.40 MPa). As shown in Fig. 2.46a1, the buckling pattern of the perfect model was nearly axisymmetric in the initial stages of buckling, while the first buckle in the imperfect model occurred at the location of 0° where the bulging of weld was introduced because it promoted the occurrence of the buckle 0°; see Fig. 2.46b1. As the pressure increased, for the perfect model, some "small" buckles grew large, as shown in Fig. 2.46a2, while in the imperfect model, more buckles formed near the first buckle (buckle 0°), as shown in Fig. 2.46b2. With the further increase in pressure, the buckles of the perfect model became bigger, as shown in Fig. 2.46a3; for the imperfect model, more buckles formed along the circumference of the knuckle, as shown in Fig. 2.46b3, and the buckling pattern was similar to that of the perfect model. At a relatively higher pressure, buckle height decreases, as shown in Fig. 2.46a4 and b4, which means that this buckling is self-limited for the perfect and imperfect head models.

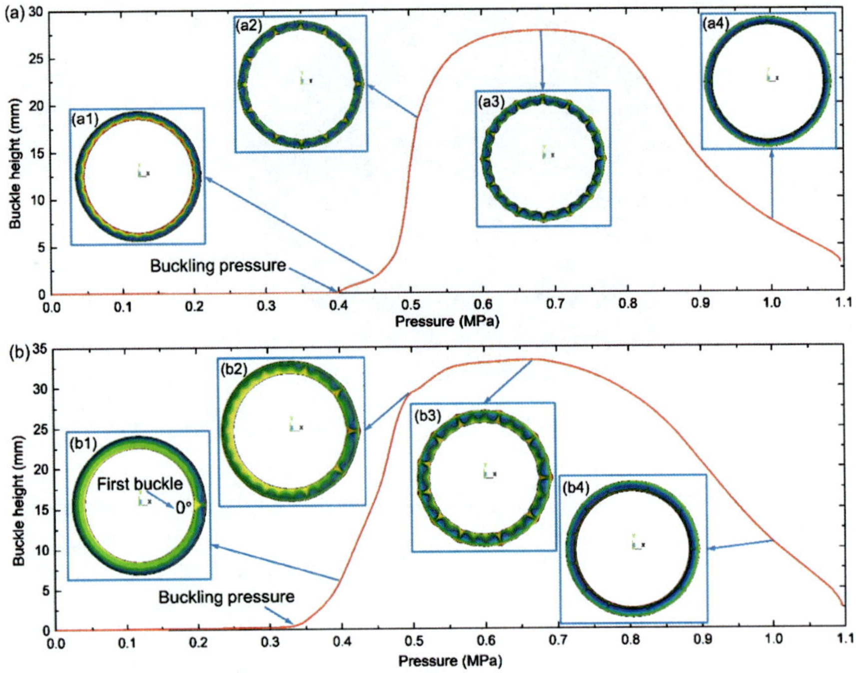

Fig. 2.46 Comparison of buckling behaviors of **a** perfect and **b** imperfect FE models (from Ref. [9] with permission of Elsevier)

2.4.4 Development of Buckling Criterion

The buckling criterion is intended to determine whether ellipsoidal heads buckle or not. First, FE models of ellipsoidal heads with cylinders were used to develop the buckling criterion. As concluded in Sect. 2.4.3.1, the strain hardening of material can increase buckling pressure and even cause the occurrence of no buckling, but it has little or no effect on thinner heads. As strain hardening cannot lead to unconservative predictions, it is omitted here for simplicity. In addition, the effects of Young's modulus and Poisson's ratio are not considered, because plastic buckling occurs in ellipsoidal heads. Yield strength is a more critical factor affecting plastic buckling pressure. Thus, material behavior in this section is assumed to be elastic-perfectly-plastic: Young's modulus of $E = 200$ GPa, Poisson's ratio of $\nu = 0.3$ and yield strength of $S_y = 100$, 200, 400 and 600 MPa.

As discussed in Sects. 2.4.3.2 and 2.4.3.4, as the diameter-to-thickness ratio and/or radius-to-height ratio decrease, no buckling occurs in ellipsoidal heads under internal pressure. A series of FE models with different diameter-to-thickness ratios and radius-to-height ratios were computed to determine whether buckling occurs in head models. Figure 2.47 gives the FE results on the diameter-to-thickness ratios and radius-to-height ratios of head models with and without buckling. The buckling

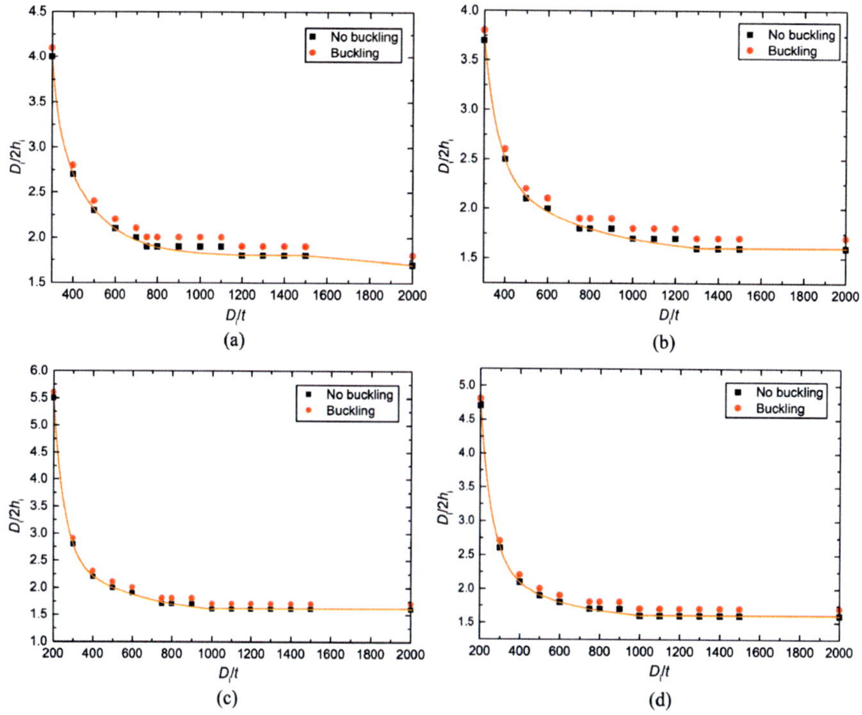

Fig. 2.47 Buckling boundaries based on FE results for different yield strengths of **a** 100 MPa, **b** 200 MPa, **c** 400 MPa and **d** 600 MPa (from Ref. [10])

boundary is the integration of diameter-to-thickness ratios and radius-to-height ratios of head models with no buckles. Figure 2.48 shows the buckling boundaries based on FE results for different yield strengths.

To modify the buckling boundaries based on FE results, we summarized the experimental results of the buckling of 23 ellipsoidal heads covering different geometrical parameters, material types and methods of fabrication, as shown in Table 2.7, together with experimental buckling pressures. For six large-scale ellipsoidal heads assembled from formed segments presented in Sect. 2.2.1.1, buckling occurred in the ZJU-FA1 and ZJU-FA2 heads with D_i/t of 872 and $D_i/2h_i$ of 1.728 and the ZJU-FT1 and ZJU-FT2 heads with D_i/t of 909 and $D_i/2h_i$ of 2, but no buckling occurred in the ZJU-FN1 and ZJU-FN2 heads with D_i/t of 909 and $D_i/2h_i$ of 1.591. In addition, we conducted failure experiments on a series of ellipsoidal heads (ZJU heads) fabricated by cold pressing and cold spinning (see Sect. 3.2), and buckling occurred at eleven heads, as shown in Table 2.7. These heads have two types of diameters: 1200 and 1800 mm. Seven ellipsoidal heads with a radius-to-height ratio of 2 were formed by cold pressing, while the others which have radius-to-height ratios of 2.2–2.4 were fabricated by cold spinning. The yield strength and tensile strength of the material were measured by tensile tests for the stainless steel plates used in these ellipsoidal

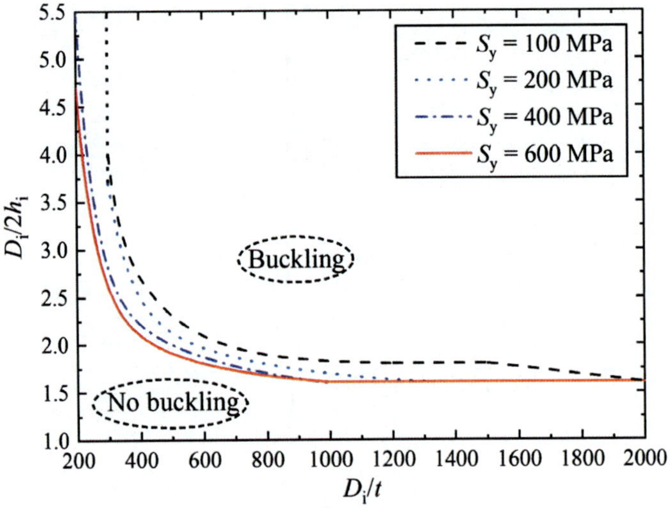

Fig. 2.48 Buckling boundaries based on FE results (from Ref. [10])

heads. For these ellipsoidal heads formed by cold pressing and cold spinning, the experimental buckling pressure was determined by the occurrence of the first buckle using video monitoring; some typical buckles are presented in Fig. 2.49.

As reported by Roche and Alix [29], the buckling tests of 17 ellipsoidal heads with a diameter of 500 mm were performed. The tested heads made of aluminium-magnesium alloy are not considered in our study. Moreover, the tested heads with a radius-to-height ratio of 5, which is not common in engineering. In addition, the mechanical property was not measured for one head made of carbon steel. Therefore, the experimental data is only available for four of 17 tested ellipsoidal heads reported by Roche and Alix, and the four heads are marked by RA1, RA2, RA3 and RA4. Galletly [30] conducted the internally pressurized buckling tests on two 2:1 ellipsoidal heads made of mild steel. The two heads are marked as Gal1 and Gal2. They are 137 mm in diameter but have very small thickness of 0.127 mm. Both of them were machined from solid billets. However, ellipsoidal heads used for pressure vessels are not often machined from solid billets. In addition, the two torispherical heads tested by Miller et al. [15] are approximate 2:1 ellipsoidal heads (Heads Nos. CBI1 and CBI2) assembled from formed segments, and their geometric and material parameters and experimental buckling pressure are shown in Table 2.7.

The buckling boundaries were modified by the experimental results of the ellipsoidal heads to consider the effects due to head fabrication such as shape imperfection, as shown in Fig. 2.50. We can see that the modified buckling boundary for yield strength of 600 MPa is in better agreement with experimental results, including those of the ZJU-FN1 and ZJU-FN2 heads with no buckling. Therefore, the modified buckling boundary for yield strength of 600 MPa was developed as the buckling criterion for ellipsoidal heads, as shown in Fig. 2.51. Based on the buckling criterion shown in

Table 2.7 Experimental results on buckling of ellipsoidal heads under internal pressure (from Ref. [5] with permission of Elsevier)

Head no	Nominal dimensions			Material	Yield strength S_y (MPa)	Tensile strength S_u (MPa)	Method of fabrication	Experimental buckling pressure, P_{exp} (MPa)
	D_i(mm)	$\frac{D_i}{t}$	$\frac{D_i}{2h_i}$					
ZJU-P22	1800	600	2.0	S30408	385	689	Cold pressing	1.30
ZJU-P20	1800	600	2.0	S30408	385	689		1.68
ZJU-P18	1800	600	2.0	S31603	310	562		1.43
ZJU-P24	1200	480	2.0	SUS3042B	444	775		1.81
ZJU-P25	1200	480	2.0	SUS3042B	444	775		2.0
ZJU-P21	1200	400	2.0	S30408	385	689		3.63
ZJU-P12	1200	480	2.0	316L2B	335	623		2.07
ZJU-S3	1800	450	2.2	S30408	392	708	Cold spinning	2.51
ZJU-S5	1800	450	2.4	S30408	392	708		1.77
ZJU-S6	1200	400	2.4	S31603	340	605		1.47
ZJU-S4	1200	400	2.4	S30408	327	700		1.45
RA1	500	500	2.5	Carbon steel	310	340	Cold spinning	0.80
RA2	500	250	2.5	Carbon steel	300	330		2.50
RA3	500	500	2.5	Stainless steel	480	690		1.50
RA4	500	250	2.5	Stainless steel	520	700		4.90
Gal1	137	1080	2	Mild steel	242	-	Machined from solid billet	0.83
Gal2	137	1080	2	Mild steel	242	-		0.76

(continued)

Table 2.7 (continued)

Head no	Nominal dimensions			Material	Yield strength S_y (MPa)	Tensile strength S_u (MPa)	Method of fabrication	Experimental buckling pressure, P_{exp} (MPa)
	D_i(mm)	$\frac{D_i}{t}$	$\frac{D_i}{2h_i}$					
ZJU-FA1	4797	872	1.728	SA738 Gr. B	509	602	Assembled from formed segments	1.15
ZJU-FA2	4797	872	1.728	SA738 Gr. B	588	662		1.15
ZJU-FT1	5000	909	2	SA738 Gr. B	612	674		0.70
ZJU-FT2	5000	909	2	SA738 Gr. B	558	626		0.70
CBI1	4877	1024	2	SA516 Gr. 70	353	552		0.40
CBI2	4877	768	2	SA516 Gr. 70	361	538		0.73

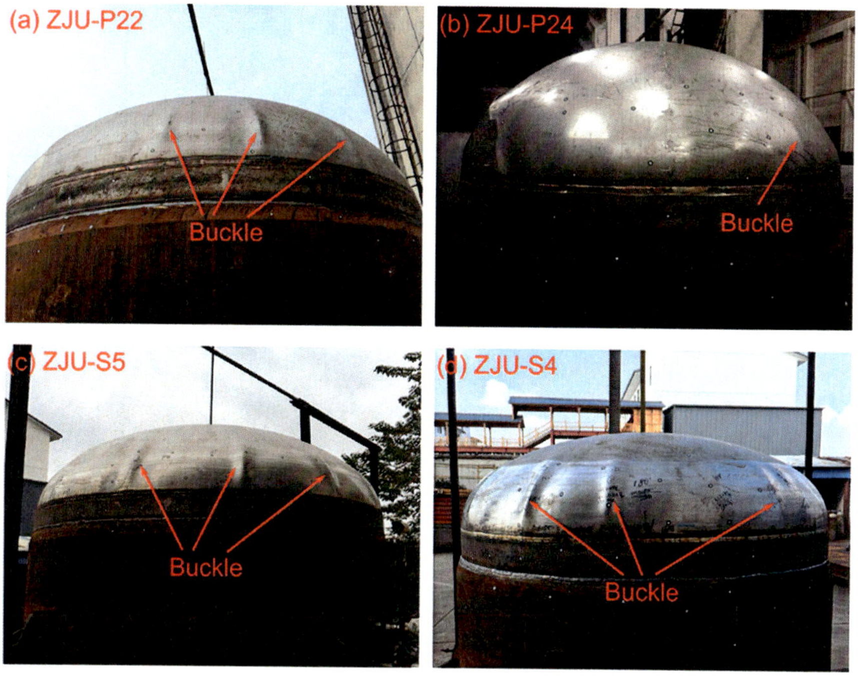

Fig. 2.49 Typical buckles in ellipsoidal heads fabricated by cold pressing and cold spinning (from Ref. [5] with permission of Elsevier)

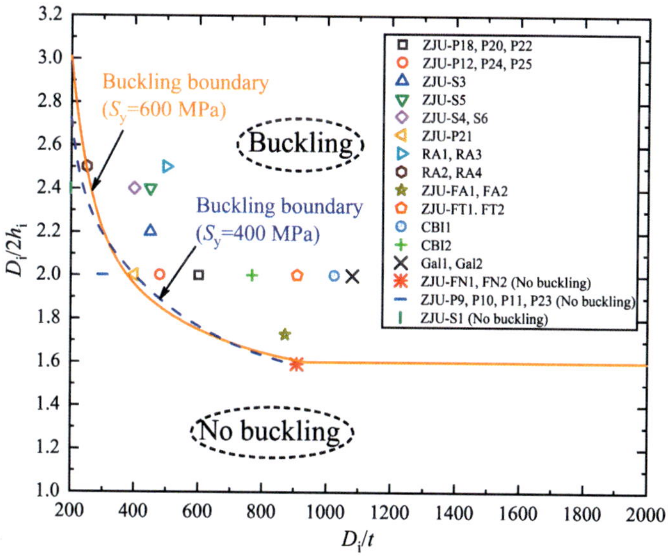

Fig. 2.50 Buckling boundaries modified by experiments (from Ref. [10])

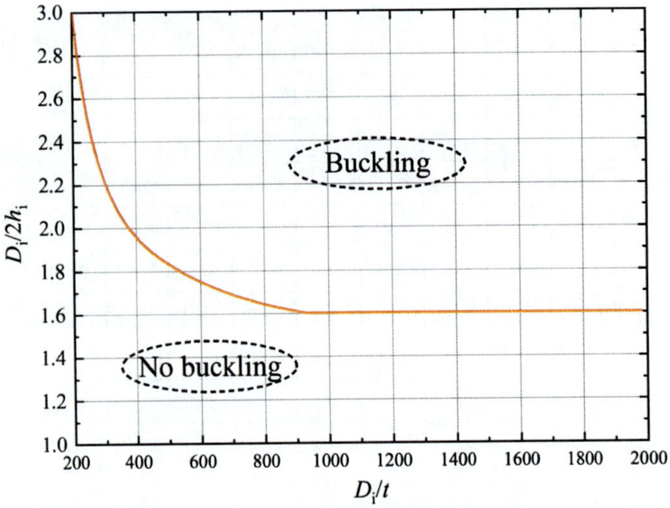

Fig. 2.51 Buckling criterion for ellipsoidal heads under internal pressure (from Ref. [10])

Fig. 2.51, we can determine whether or not an internally pressurized ellipsoidal head will buckle, depending on the ratios of diameter-to-thickness and radius-to-height. If the ratios of diameter-to-thickness and radius-to-height are located in the curve or top-right corner, as shown in Fig. 2.51, ellipsoidal heads with these ratios will buckle. No buckling will occur in ellipsoidal heads when these ratios are located in bottom-left corner, as shown in Fig. 2.51.

2.5 Formula for Predicting Buckling Pressure

2.5.1 Parameter Study

We carried out a parametric study to obtain a formula for predicting the buckling pressure of ellipsoidal heads. As mentioned in Sects. 2.4.3.2 and 2.4.3.4, the diameter-to-thickness ratio and radius-to-height ratio of an ellipsoidal head significantly affect buckling pressure. Thus, these two geometric parameters (diameter-to-thickness ratio and radius-to-height ratio) need to be considered in formulas for predicting buckling pressure. The values of the two geometric parameters are determined as follows: $D_i/2h_i = 1.7$, 2.0, 2.2, 2.6 and 3.0, and $D_i/t = 300$, 400, 500, 750, 1000, 1500 and 2000.

As discussed in Sect. 2.4.3.1, the strain hardening of material can either increase buckling pressure or even result in no buckling at all, but it has little or no effect on thinner heads. Thus, strain hardening of material is excluded here for simplicity as this cannot lead to non-conservative predictions. Compared with Young's modulus,

Poisson's ratio and the strain hardening of material, yield strength is a critical factor affecting plastic buckling pressure. Thus, the material behavior in this section is assumed to be elastic-perfectly-plastic: Young's modulus of $E = 200$ GPa, Poisson's ratio of $\nu = 0.3$, and yield strength values of $S_y = 100$, 200, 400 and 600 MPa.

As discussed in Sect. 2.4.3.3, buckling pressure of an ellipsoidal head with a cylinder is generally lower than that of a head without a cylinder, but a cylinder has little or no effect when ellipsoidal heads become very thin and/or very high. Thus, a cylinder, commonly attached to an ellipsoidal head, is considered in FE models herein; this cannot lead to un-conservative predictions.

Therefore, the buckling pressures of perfect ellipsoidal heads with cylinders made of elastic-perfectly-plastic material are calculated for the above values of diameter-to-thickness ratio, radius-to-height ratio and yield strength ($D_i/2h_i = 1.7, 2.0, 2.2, 2.6$ and 3.0, $D_i/t = 300, 400, 500, 750, 1000, 1500$ and 2000, and $S_y = 100, 200, 400$ and 600 MPa), and the calculated buckling pressures are shown in Table 2.8. It can be seen that a total of 95 ellipsoidal head cases have buckling pressure, while no buckling occurs at some ellipsoidal heads with smaller diameter-to-thickness ratios and/or smaller radius-to-height ratios, as discussed in Sect. 2.4.4.

We further investigated the relationships between P_b/S_y and D_i/t or $D_i/2h_i$ based on the FE results shown in Table 2.8. For different cases of $D_i/2h_i$, the relationships between P_b/S_y and D_i/t are shown in Fig. 2.52. We can see that P_b/S_y shows an exponential relationship with D_i/t, and the effect of yield strength (S_y) is small. As mentioned in Sect. 1.3.3, Galletly Formula (1.20) for plastic buckling pressure is also expressed by an exponential relationship of P_b/S_y and D_i/t. Similar to the relationship of P_b/S_y and D_i/t, P_b/S_y shows an approximately exponential relationship with $D_i/2h_i$ for different cases of D_i/t (Fig. 2.53).

2.5.2 *Development of New Formula*

2.5.2.1 New Formula for Perfect Ellipsoidal Heads

In Figs. 2.52 and 2.53, P_b/S_y shows an exponential relationship with diameter-to-thickness ratio (D_i/t) and radius-to-height ratio ($D_i/2h_i$). Thus, a formula was generated by nonlinear curve fitting (Fig. 2.54). The statistics of this fitting are as follows: the coefficient of determination (adjusted R-Square) is 0.977, and the residual sum of squares is 2.95×10^{-6}. In addition, 95% confidence and 95% prediction bands for the FE results are shown in Fig. 2.54. Therefore, a new formula for the buckling pressure of a perfect ellipsoidal head with elastic-perfectly-plastic material is proposed as follows:

$$P_b = 56S_y\left(\frac{t}{D_i}\right)^{1.29}\left(\frac{2h_i}{D_i}\right)^{1.93} \tag{2.1}$$

Table 2.8 Buckling pressures calculated by FE models (from Ref. [5] with permission of Elsevier)

S_y (MPa)	D_i/t	Buckling pressure, P_b (MPa)				
		$D_i/2h_i =$ 1.7	$D_i/2h_i =$ 2.0	$D_i/2h_i =$ 2.2	$D_i/2h_i =$ 2.6	$D_i/2h_i =$ 3.0
100	400	NB	NB	NB	NB	0.33
	500	NB	NB	NB	0.30	0.23
	750	NB	0.31	0.27	0.18	0.13
	1000	NB	0.22	0.17	0.11	0.091
	1500	NB	0.128	0.105	0.0705	0.0536
	2000	0.14	0.086	0.0687	0.0489	0.0368
200	400	NB	NB	NB	0.79	0.62
	500	NB	NB	0.79	0.63	0.46
	750	NB	0.62	0.47	0.33	0.26
	1000	NB	0.40	0.31	0.22	0.17
	1500	0.37	0.22	0.18	0.13	0.11
	2000	0.27	0.18	0.13	0.093	0.071
400	300	NB	NB	NB	NB	1.90
	400	NB	NB	NB	1.60	1.21
	500	NB	NB	1.60	1.13	0.86
	750	NB	1.07	0.84	0.62	0.48
	1000	1.18	0.74	0.57	0.42	0.32
	1500	0.75	0.48	0.36	0.26	0.19
	2000	0.563	0.355	0.275	0.185	0.134
600	300	NB	NB	NB	NB	2.84
	400	NB	NB	3.20	2.24	1.72
	500	NB	2.72	2.14	1.58	1.24
	750	NB	1.56	1.22	0.85	0.67
	1000	1.79	1.16	0.88	0.60	0.46
	1500	1.17	0.74	0.57	0.39	0.28
	2000	0.87	0.57	0.44	0.271	0.210

NB = No buckling

2.5.2.2 New Formula for Actual Ellipsoidal Heads

As mentioned in Sect. 2.4.3.5, the buckling pressure of an ellipsoidal head is sensitive to shape imperfection. In order to develop a new formula for the buckling pressure of actual ellipsoidal heads, Formula (2.1) for perfect ellipsoidal heads is modified by reduction factor (β) considering initial shape imperfection, as given by

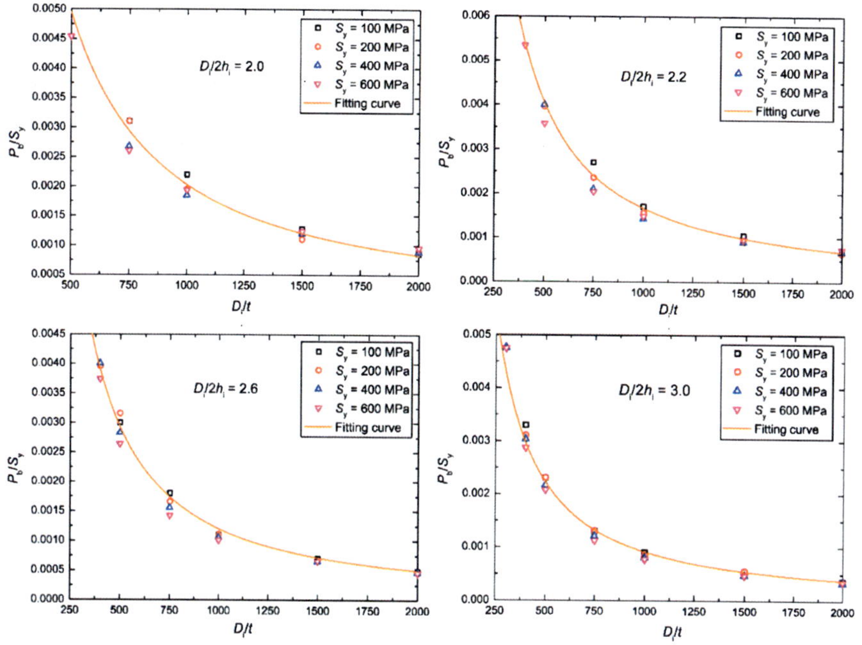

Fig. 2.52 Relationships of P_b/S_y and D_i/t for different cases of $D_i/2h_i$ (from Ref. [5] with permission of Elsevier)

$$P_b = 56\beta S_y \left(\frac{t}{D_i}\right)^{1.29} \left(\frac{2h_i}{D_i}\right)^{1.93} \tag{2.2}$$

where $\beta = 1.0$ for steel ellipsoidal heads fabricated by cold pressing and spinning, $\beta = 0.625$ for steel ellipsoidal heads assembled from formed segments.

Shape imperfection is associated with the fabrication methods of ellipsoidal heads. Steel ellipsoidal heads fabricated by cold pressing and spinning have small shape imperfections. The predictions of Formula (2.2) for steel ellipsoidal heads fabricated by cold pressing and spinning are in relatively good agreement with the experimental results (Fig. 2.55). Thus, $\beta = 1.0$ for steel ellipsoidal heads fabricated by cold pressing and spinning. However, for steel ellipsoidal heads assembled from formed segments, $\beta = 0.625$, a difference that accounts for the large initial shape imperfections. Figure 2.55 shows that the predictions of new Formula (2.2) are in reasonably good agreement with the experimental results of ZJU-FA, ZJU-FT and CBI heads assembled from formed segments. Therefore, the new Formula (2.2) was developed to predict the buckling pressure of actual ellipsoidal heads.

Formula (2.2) is developed on basis of the FE results of perfect ellipsoidal heads with $D_i/2h_i$ of 1.7–3.0 and D_i/t of 300–2000 because no buckling occurs in perfect heads with $1.6 \leq D_i/2h_i < 1.7$ or $200 \leq D_i/t < 300$. However, bucking may occur in actual heads with $1.6 \leq D_i/2h_i < 1.7$ or $200 \leq D_i/t < 300$ according to the new

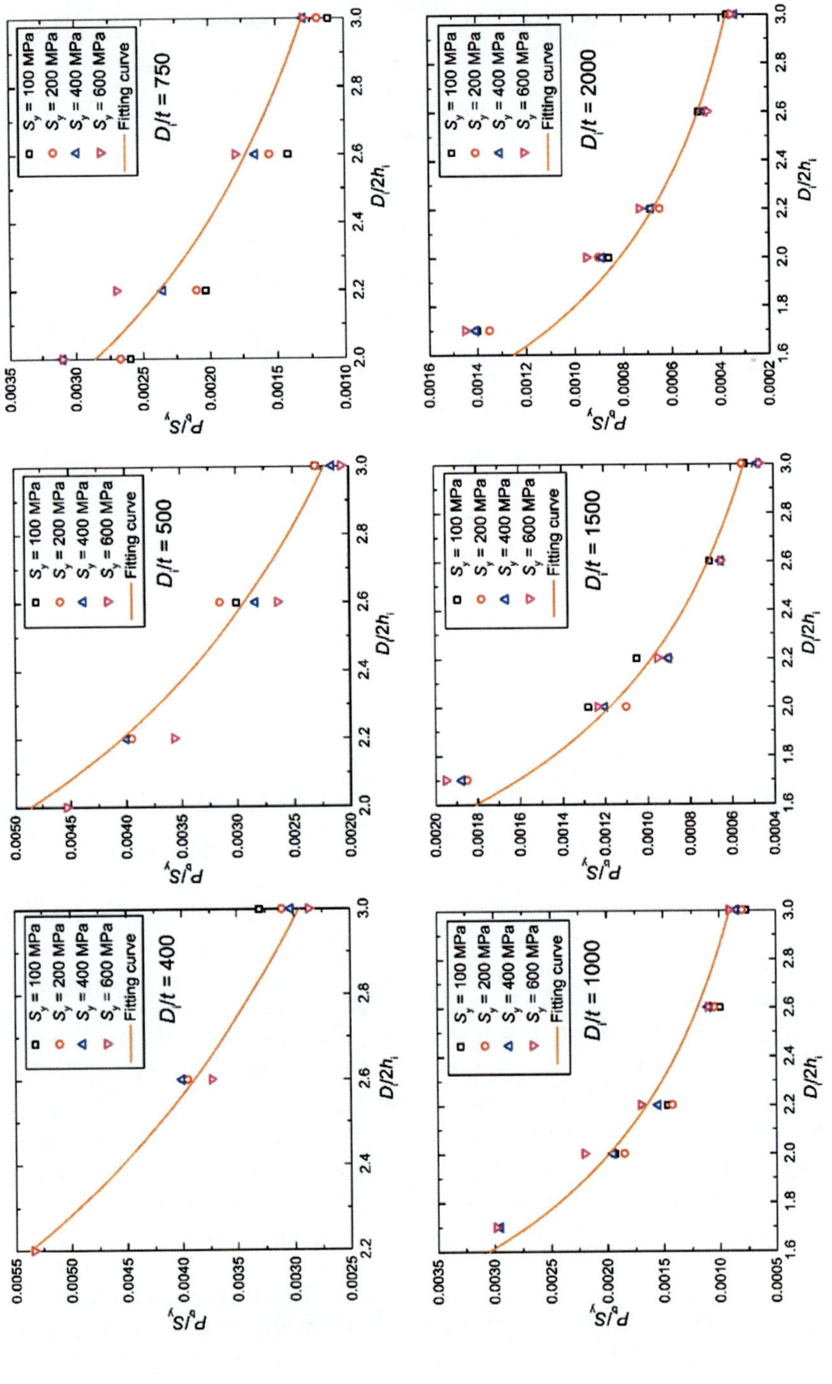

Fig. 2.53 Relationships of P_b/S_y and $D_i/2h_i$ for different cases of D_i/t (from Ref. [5] with permission of Elsevier)

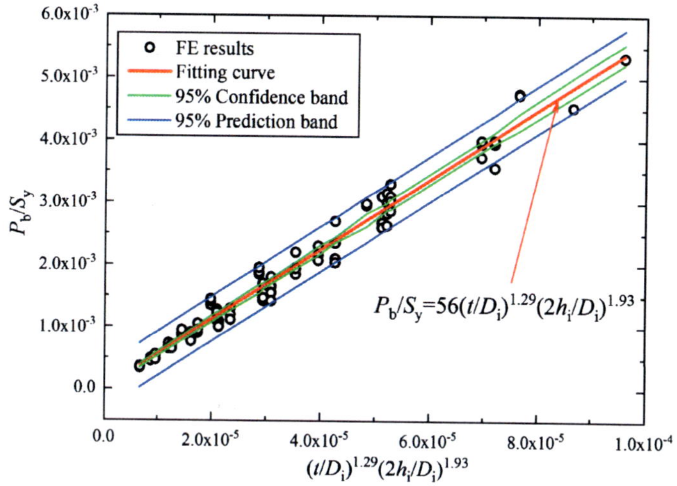

Fig. 2.54 Curve fitting for FE results of buckling pressure (from Ref. [5] with permission of Elsevier)

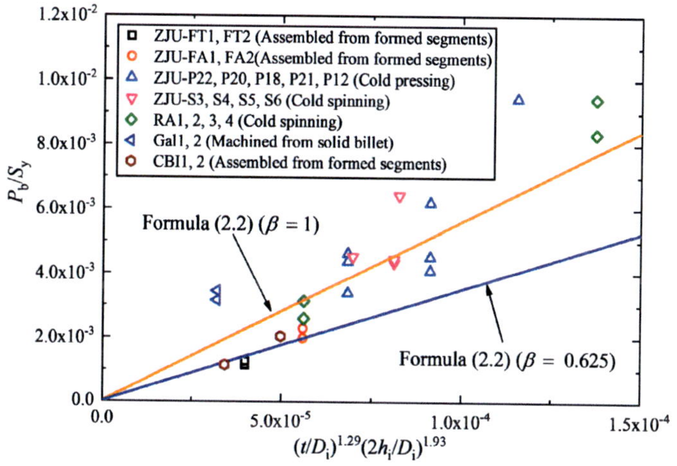

Fig. 2.55 Comparison of new Formula (2.2) and experimental results (from Ref. [5] with permission of Elsevier)

buckling criteria (Fig. 2.51). Thus, Formula (2.2) is suggested to be applicable to ellipsoidal heads with $D_i/2h_i$ of 1.6–3.0 and D_i/t of 200–2000. We can see from Table 2.10 that the predictions of Formula (2.2) are in good agreement with the experimental buckling pressures of the RA3 and RA4 heads with D_i/t of 250. But it should be noted that Formula (2.2) for plastic buckling pressure is not applicable for heads with $D_i/2h_i > 3.0$ or $D_i/t > 2000$ because elastic buckling may occur in such heads.

2.5.3 *Comparison Between New Formula and Existing Formulas*

The existing formulas for predicting the buckling pressure of ellipsoidal heads under internal pressure are reviewed in Sect. 1.3.3. Table 2.9 lists six formulas for predicting the buckling pressure of ellipsoidal heads. Tovstik Formula (1.18) is applicable to ellipsoidal heads with a small height ($D_i/2h_i \geq 3.3$) and is not included in the summary of experimental results given here; hence, Table 2.10 does not include a comparison between the experimental results and the corresponding predictions of Tovstik Formula (1.18). In fact, ellipsoidal heads with such small heights are not common in engineering application. Two Formulas (1.19) and (1.20) proposed by Galletly et al. are only applicable to 2:1 ellipsoidal heads. Galletly Formula (1.19) for elastic buckling pressure is applicable to 2:1 ellipsoidal heads with a diameter-to-thickness ratio of 750–1500, while Galletly Formula (1.20) for plastic buckling pressure is applicable to 2:1 ellipsoidal heads with a diameter-to-thickness ratio of 600–1500. Roche and Alix did not discuss the limitations of Formulas (1.21) and (1.22). Roche and Alix Formula (1.22) for plastic buckling pressure is an empirical expression determined by experimental results in which the tested heads include two radius-to-height ratios ($D_i/2h_i$) of 2.5 and 5, and three diameter-to-thickness ratios (D_i/t) of 250, 500 and 1000. With respect to the limitations of $D_i/2h_i$ and D_i/t for ellipsoidal heads that are usually used in engineering, the new Formula (2.2) proposed in this book has good applicability in most situation.

Table 2.10 and Fig. 2.56 present a comparison between the experimental data and predicted buckling pressures calculated by the formulas for buckling pressure. The ratios of the predicted and experimental buckling pressures (P_b/P_{exp}) and Root Mean Square Error (RMSE) are used to evaluate the accuracy of the prediction formulas. The smaller the RMSE, the closer predictions of the formulas are to the experimental results.

Formulas (1.19) and (1.21) are used to predict elastic buckling pressure, whereas Formulas (1.20), (1.22) and (2.2) are used to predict plastic buckling pressure. The mean values of the ratios P_b/P_{exp} (mean of P_b/P_{exp}) and RMSE for Formulas (1.19) and (1.21) are much higher than those of Formulas (1.20), (1.22) and (2.2), and the mean values of the ratios P_b/P_{exp} for Formulas (1.20), (1.22) and (2.2) are close to 1, demonstrating that Formulas (1.20), (1.22) and (2.2) provide much more accurate predictions than Formulas (1.19) and (1.21), as shown in Table 2.10 and Fig. 2.56. Since plastic buckling occurs in ellipsoidal heads with the parameters ($D_i/2h_i$, D_i/t and S_y) investigated in this book, the plastic buckling predictions of Formulas (1.21), (1.22) and (2.2) are more accurate than the elastic buckling predictions of Formulas (1.19) and (1.21).

Table 2.10 and Fig. 2.56 also show that the mean of P_b/P_{exp} and RMSE for new Formulas (2.2) are smaller than those for empirical Roche and Alix Formula (1.22), indicating that Formula (2.2) more accurately predicts the experimental data. Moreover, compared with empirical Formula (1.22), Formula (2.2) is a simpler expression, which is an asset for engineering applications.

Table 2.9 Summary of formulas for predicting buckling pressure of ellipsoidal heads under internal pressure (from Ref. [5] with permission of Elsevier)

Formula no	Authors	Prediction formula	Limitation	Buckling type
(1.18)	Tovstik	$P_b = \frac{16E}{\sqrt{3(1-\nu^2)}}\left(\frac{b/a}{a/t}\right)^2\left[1+\sqrt{\frac{193}{32\sqrt{3(1-\nu^2)}}}\sqrt{\frac{1-4(b/a)^2}{b/t}}\right]$	$D_i/2h_i \geq 3.3$	Elastic buckling
(1.19)	Galletly	$P_b = 455E(D_i/t)^{-2.5}$	$750 < D_i/t < 1500$; $D_i/2h_i = 2$	Elastic buckling
(1.20)	Galletly	$P_b = 10.4S_y(D_i/t)^{-1.25}$	$600 < D_i/t < 1500$; $D_i/2h_i = 2$	Plastic buckling
(1.21)	Roche and Alix	$P_b = K_e\frac{\pi^2 E}{1-\nu^2}\frac{t^2 h_i^2}{(D_i/2)^2}$	–	Elastic buckling
(1.22)	Roche and Alix	$P_b = K_c S_f\frac{t^{5/3}h_i^{4/3}}{(D_i/2)^3}$	–	Plastic buckling
(2.2)	New formula (Zheng and Li)	$P_b = 56\beta S_y\left(\frac{t}{D_i}\right)^{1.29}\left(\frac{2h_i}{D_i}\right)^{1.93}$	$200 \leq D_i/t \leq 2000$; $1.6 \leq D_i/2h_i \leq 3$	Plastic buckling

Table 2.10 Comparison between buckling pressures calculated by formulas and experimental data (from Ref. [5] with permission of Elsevier)

Head no	Experiment, P_{exp} (MPa)	Galletly Formula (1.19)		Roche and Alix Formula (1.21)		Galletly Formula (1.20)		Roche and Alix Formula (1.22)		New Formula (2.2)	
		P_b(MPa)	$\frac{P_b}{P_{exp}}$	P_b(MPa)	$\frac{P_b}{P_{exp}}$	P_b(MPa)	$\frac{P_b}{P_{exp}}$	P_b(MPa)	$\frac{P_b}{P_{exp}}$	P_b(MPa)	$\frac{P_b}{P_{exp}}$
ZJU-P22	1.3	–	–	5.87	4.52	1.35	1.04	2.21	1.70	1.48	1.14
ZJU-P20	1.68	–	–	5.87	3.49	1.35	0.80	2.21	1.32	1.48	0.88
ZJU-P18	1.43	–	–	5.87	4.10	1.09	0.76	1.80	1.26	1.19	0.83
ZJU-P24	1.81	–	–	9.18	5.07	–	–	3.65	2.02	2.27	1.25
ZJU-P25	2.0	–	–	9.18	4.59	–	–	3.65	1.83	2.27	1.14
ZJU-P21	3.63	–	–	13.22	3.64	–	–	4.35	1.20	2.49	0.69
ZJU-P12	2.07	–	–	9.18	4.43	–	–	2.86	1.38	1.71	0.83
ZJU-S3	2.51	–	–	8.63	3.44	–	–	2.90	1.16	1.81	0.72
ZJU-S5	1.77	–	–	7.25	4.10	–	–	2.40	1.36	1.53	0.86
ZJU-S6	1.47	–	–	9.18	6.24	–	–	2.51	1.71	1.55	1.05
ZJU-S4	1.45	–	–	9.18	6.33	–	–	2.73	1.88	1.49	1.03
RA1	0.8	–	–	5.19	6.49	–	–	1.10	1.38	0.98	1.23
RA2	2.5	–	–	19.12	7.65	–	–	3.38	1.35	2.31	0.92
RA3	1.5	–	–	4.78	3.19	–	–	1.98	1.32	1.51	1.01
RA4	4.9	–	–	19.94	4.07	–	–	6.55	1.34	4.01	0.82
Gal1	0.83	2.34	2.82	1.83	2.20	0.41	0.49	0.38	0.46	0.44	0.53
Gal2	0.76	2.34	3.08	1.83	2.41	0.41	0.54	0.38	0.50	0.44	0.58
ZJU-FA1	1.15	–	–	5.17	4.50	–	–	1.97	1.71	1.14	0.99
ZJU-FA2	1.15	–	–	4.63	4.03	–	–	2.21	1.92	1.32	1.15

(continued)

Table 2.10 (continued)

Head no	Experiment, P_{exp} (MPa)	Galletly Formula (1.19)		Roche and Alix Formula (1.21)		Galletly Formula (1.20)		Roche and Alix Formula (1.22)		New Formula (2.2)	
		P_b(MPa)	$\frac{P_b}{P_{exp}}$	P_b(MPa)	$\frac{P_b}{P_{exp}}$	P_b(MPa)	$\frac{P_b}{P_{exp}}$	P_b(MPa)	$\frac{P_b}{P_{exp}}$	P_b(MPa)	$\frac{P_b}{P_{exp}}$
ZJU-FT1	0.7	3.65	5.21	2.62	3.74	1.28	1.83	1.33	1.90	0.98	1.40
ZJU-FT2	0.7	3.68	5.26	2.64	3.77	1.16	1.66	1.22	1.74	0.89	1.27
CBI1	0.4	2.57	6.43	1.96	4.90	0.63	1.58	0.78	1.95	0.48	1.20
CBI2	0.73	5.57	7.63	3.68	5.04	0.93	1.27	1.24	1.70	0.72	0.99
Mean	–	–	5.07	–	4.43	–	1.11	–	1.48	–	0.98
RMSE (MPa)	–	2.25		6.84		0.36		0.90		0.40	

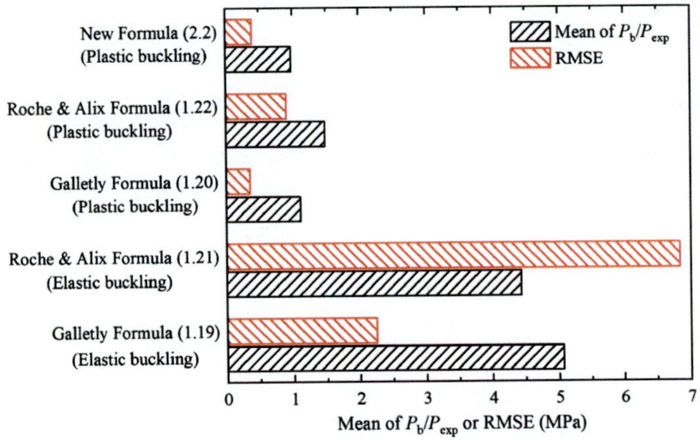

Fig. 2.56 Comparison of different formulas to predict buckling pressure of ellipsoidal heads (from Ref. [5] with permission of Elsevier)

The Galletly Formula (1.20) also provides good prediction for eleven out of the twenty-three experimental results. However, Galletly Formula (1.20) is only applicable to an ellipsoidal head with a radius-to-height ratio of 2. Compared to Formula (1.20), the new Formula (2.2) is applicable to wider ranges of radius-to-height ratios and diameter-to-thickness ratios.

For perfect ellipsoidal heads with a radius-to-height ratio of 2, the new Formula (2.2) can be transformed into Formula (2.3), thus:

$$P_b = 14.7S_y(D_i/t)^{-1.29} \tag{2.3}$$

By comparing Formula (2.3) with Galletly Formula (1.20), we can analyze the ratio of the buckling pressures determined by these two formulas; that is, a ratio of $0.7(D_i/t)^{0.04}$. When D_i/t varies from 600 to 1500, the range of this ratio is 0.90–0.94, indicating that Formula (2.3) is in good agreement with Galletly Formula (1.20). As mentioned in Sect. 1.3.3, Formula (1.20) is derived from the numerical data obtained by the BOSOR 5 program, where buckling pressure is computed using pre-buckling and plastic bifurcation analyses based on a finite difference method. Formula (2.2) is derived from the numerical data obtained by finite element analysis using the arc-length method. Although these two formulas are based on different numerical methods, they are in good agreement with respect to ellipsoidal heads with a radius-to-height ratio of 2. Thus, the two formulas verify each other.

In summary, a new formula for predicting the buckling pressure of a perfect ellipsoidal head is proposed based on the parameter study of FE method. Building on the formula for perfect ellipsoidal heads, a new formula is developed for predicting buckling pressure of actual ellipsoidal heads after considering a reduction factor that accounts for the effect of initial shape imperfection, and depends on the methods of head fabrication, i.e. whether cold pressing and spinning or assembly by formed segments involved in construction. The predictions of the new Formula (2.2) for actual ellipsoidal heads are in reasonably good agreement with the experimental results. Compared with other formulas, the new Formula (2.2) has a comprehensive advantage in both accuracy and applicability.

References

1. Eslami RM (2015) Buckling and postbuckling of beams, plates, and shells. Springer International Publishing AG, Cham
2. Bushnell D (1981) Buckling of shells—pitfall for designers. AIAA J 19:1183–1226
3. Bushnell D (1989) Computerized buckling analysis of shells. Kluwer Academic Publishers, Dordrecht
4. Singer J, Arbocz J, Weller T (1998) Buckling experiments: experimental methods in buckling of thin-walled structures. Wiley, New York
5. Li K, Zheng J, Zhang Z, Gu C, Xu P, Chen Z (2019) A new formula to predict buckling pressure of steel ellipsoidal head under internal pressure. Thin-Walled Struct 144:106311
6. Zhang X (2016) Research on bearing capacity testing machine for nuclear power plants steel containment ellipsoidal heads. Master's Thesis, Zhejiang University, Hangzhou (in Chinese)
7. Li K, Zheng J, Zhang Z, Gu C, Zhang X, Liu S, Ge H, Gu C, Lin G (2017) Experimental investigation on buckling of ellipsoidal head of steel nuclear containment. ASME J Pressure Vessel Technol 139:061206
8. Zheng J, Li K, Liu S, Ge H, Zhang Z, Gu C, Qian H, Hua Z (2018) Effect of shape imperfection on the buckling of large-scale thin-walled ellipsoidal head in steel nuclear containment. Thin-Walled Struct 124:514–522
9. Li K, Zheng J, Liu S, Ge H, Sun G, Zhang Z, Gu C, Xu P (2019) Buckling behavior of large-scale thin-walled ellipsoidal head under internal pressure. Thin-Walled Struct 141:260–274
10. Li K (2019) Research on buckling behavior and prediction method of large-scale thin-walled ellipsoidal head under internal pressure. PhD Thesis, Zhejiang University, Hangzhou (in Chinese)
11. Committee ASTM (2020) ASTM A370–20: Standard test methods and definitions for mechanical testing of steel products. ASTM, West Conshohocken
12. Liu P, Zheng J, Ma L, Miao C, Wu L (2008) Calculations of plastic collapse load of pressure vessel using FEA. J Zhejiang Univ Sci A 9(7):900–906
13. Committee ANS (2019) ASME BPVC II-D: ASME boiler and pressure vessel code, section II, part D, properties (Metric). Am Soc Mechan Eng, New York
14. General Administration of Quality Supervision, Inspection and Quarantine of the People's Republic of China, Standardization Administration of the People's Republic of China (2012) GB/T 150.2–2011: Pressure vessels—part 2: materials. Standards Press of China, Beijing
15. Miller CD, Grove RB, Bennett JG (1986) Pressure testing of large-scale torispherical heads subjected to knuckle buckling. Int J Pres Ves Piping 22:147–159
16. Kainat M, Cheng JJR, Martens M, Adeeb S (2016) Measurement and characterization of the initial geometric imperfections in high strength U-ing, O-ing and Expanding manufactured steel pipes. ASME J Pressure Vessel Technol 138:021201
17. Zhang J, Zhang M, Tang W, Wang W, Wang M (2017) Buckling of spherical shells subjected to external pressure: a comparison of experimental and theoretical data. Thin-Walled Struct 111:58–64
18. Wang H, Yao X, Li L, Sang Z, Krakauer BW (2017) Imperfection sensitivity of externally-pressurized, thin-walled, torispherical-head buckling. Thin-Walled Struct 113:104–110
19. Wang B, Zhu S, Hao P, Bi X, Du K, Chen B, Ma X, Chao YJ (2018) Buckling of quasi-perfect cylindrical shell under axial compression: a combined experimental and numerical investigation. Int J Solids Struct 130–131:232–247

20. Riks E (1979) An incremental approach to the solution of snapping and buckling problems. Int J Solids Struct 15:529–551
21. Crisfield MA (1981) A fast incremental/iterative solution procedure that handles "snap-through." Comput Struct 13:55–62
22. Forde BWR, Stiemer SF (1987) Improved arc length orthogonality methods for nonlinear finite element analysis. Comput Struct 27:625–630
23. Galletly GD, Blachut J, Moreton DN (1990) Internally pressurized machined dome ends—a comparison of the plastic buckling prediction of the deformation and flow theories. Proc Inst Mechan Eng 204:169–186
24. Roche RL, Autrusson B (1986) Experimental tests on buckling of torispherical heads and methods of plastic bifurcation analysis. ASME J Pressure Vessel Technol 108:138–145
25. Teng JG, Zhao Y (2000) On the buckling of a pressure vessel with a conical end. Eng Fail Anal 7:261–280
26. Bushnell D (1981) Elastic-plastic buckling of internally pressurized ellipsoidal pressure vessel heads. WRC Bulletin 267
27. Galletly GD (1981) Plastic buckling of torispherical and ellipsoidal shells subjected to internal pressure. Proc Inst Mech Eng 195:329–345
28. Bushnell D (1982) Plastic buckling of various shells. ASME J Pressure Vessel Technol 104:51–72
29. Roche RL, Alix M (1980) Experimental tests on buckling of ellipsoidal vessel heads subjected to internal pressure. In: Proceeding of 4th international conference on pressure vessel technology, institution of mechanical engineers, pp 159–165
30. Galletly GD (1978) Elastic and elastic plastic buckling of internal pressure 2:1 ellipsoidal heads. ASME J Pressure Vessel Technol 100:335–343

Chapter 3
Plastic Collapse of Ellipsoidal Heads

3.1 Introduction to Plastic Collapse

Figure 3.1 shows a typical pressure-deformation (displacement, strain, etc.) curve for a pressure vessel made of a ductile metal. Initially, the deformation is smaller because the metal is in an elastic state. As pressure increases, when the metal is in a plastic state, the deformation becomes larger until the occurrence of plastic collapse (tensile plastic instability). The basic ideas about tensile plastic instability of pressure vessels are summarized by Updike and Kalnins [1, 2]: The onset of tensile plastic instability occurs when the following effects are balanced: the strengthening effects due to the strain-hardening of the metal and/or shape change and the weakening effects due to wall thinning and/or shape change. The wall thinning of a vessel is a critical weakening effect which is essential for the occurrence of tensile plastic instability, necking and eventual ductile fracture. The change in shape of the vessel's surface may be either strengthening or weakening depending on its effect on the relationship between loading and true stress in the vessel.

Plastic collapse corresponds to the maximum point in the pressure-deformation curve. Plastic collapse pressure is taken as the maximum load that a structure can carry. After plastic collapse, deformation continues but the pressure decreases slightly until a rupture occurs in the vessel, leading to the complete loss of load-carrying capacity. Although plastic collapse is reached prior to bursting, the pressure at burst is assumed to be plastic collapse pressure. This behavior of the deformation and plastic collapse is also seen in the uniaxial tensile test (Sect. 1.1.2). Necking corresponds to the part of the tensile test in which plastic collapse exists. The onset of necking is associated with the occurrence of maximum load during the tensile deformation of a ductile metal. Since plastic collapse is a failure mode for pressure vessels specified in ASME VIII-2, we will use the term "plastic collapse" herein.

The estimate of plastic collapse pressure can be obtained using elastic–plastic analysis by considering the effects of material hardening and geometric nonlinearity. Updike and Kalnins [1] developed a procedure to predict plastic collapse pressure of torispherical heads subjected to internal pressure. But this procedure includes

J. Zheng and K. Li, *New Theory and Design of Ellipsoidal Heads for Pressure Vessels*,
https://doi.org/10.1007/978-981-16-0467-6_3

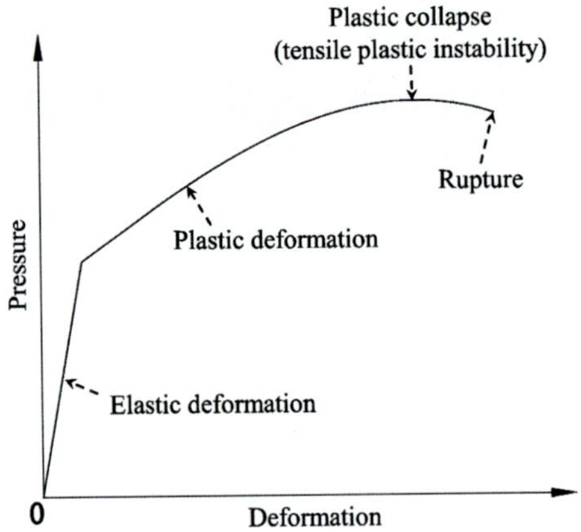

Fig. 3.1 Pressure-deformation curve of pressure vessels

many complicated equations and a computer program need be used to perform the calculation of these equations. Updike and Kalnins [2] also investigated the calculated pressure at plastic collapse of a pressure vessel, and proposed that plastic collapse pressure be accepted as an upper bound to the pressure at which a vessel ruptures. As mentioned in Sect. 2.2.1.2, Liu et al. [3] developed finite element technique using the arc-length algorithm to predict plastic collapse load of cylindrical pressure vessels, and results show that the predicted results are in good agreement with the experimental values. Kadam [4] employed the arc-length algorithm to accurately estimate static burst (plastic collapse) pressure of unflawed high pressure cylinders using nonlinear FE method, and found that the assumption of elastic-perfectly plastic material model gives a conservative prediction of burst pressure when compared to nonlinear model.

Chapter 3 is intended to provide an in-depth investigation on plastic collapse of ellipsoidal heads under internal pressure and adapted from Refs. [5–7] with permission of ASME and Elsevier.

3.2 Plastic Collapse Experiment

3.2.1 Experimental Setup

To investigate the plastic collapse of ellipsoidal heads under internal pressure, we performed plastic collapse experiments on a total of 31 ellipsoidal heads (ZJU heads) fabricated by cold pressing and cold spinning. The dimensions and materials of the tested ellipsoidal heads are given in Table 3.1. To conduct tests for the plastic collapse of the ellipsoidal heads, we constructed reusable test vessels with the same inside diameters as the tested ellipsoidal heads, as shown in Fig. 3.2. The reusable test vessels for the plastic collapse experiments of ellipsoidal heads were designed using the same method as that for the buckling experiments in Sect. 2.2.1. Each test vessel

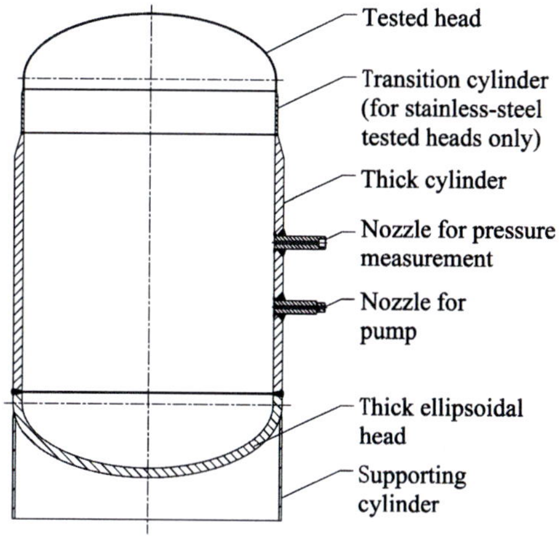

Fig. 3.2 Test vessel for plastic collapse of ellipsoidal head (from Ref. [6] with permission of Elsevier)

was mainly composed of a tested head, thick cylinder and thick 2:1 ellipsoidal head. The thick cylinders and thick ellipsoidal heads had the same thickness of 22 mm, nearly four times the maximum thickness of the tested heads. This guaranteed that even if the tested heads ruptured, the thick cylinders and thick ellipsoidal heads of the test vessels would remain in an elastic state. Thus, the thick cylinders and thick ellipsoidal heads could then be reused to weld another test head. In addition, since the stainless-steel test heads (2–3 mm) were too thin compared with the thick cylinders (22 mm), it was difficult to weld the stainless-steel test heads to cylinders of such thickness. Thus, a short cylinder was used as a transition for combining the stainless-steel test heads with the thick cylinders, as shown in Fig. 3.2. The thicknesses of the transition cylinders were twice of those of stainless-steel test heads. The length of all transition cylinders was 100 mm. The lengths of the thick cylinders for test vessels with inside diameters of 600 mm, 1200 mm and 1800 mm are 600 mm, 900 mm and 1000 mm respectively.

The test vessel was pressurized with water using a pump. A pressure sensor was used to automatically measure the inside pressure of the test vessel. Pressurization was stopped at different pressures to conduct a visual examination of the deformations of the test heads. Test vessels were pressurized until the tested ellipsoidal heads ruptured. For heads Nos. ZJU-P19 and ZJU-P23, the heads were not continuously pressurized to rupture since the pump failed to work properly.

In addition, two large approximately ellipsoidal heads (Heads Nos. CBI1 and CBI2) tested by Miller et al. [8] also ruptured, and they were assembled by formed segments. The test results of CBI1 and CBI2 heads were available for investigating plastic collapse, as shown in Table 3.1.

Table 3.1 Comparison of experimental results and plastic collapse pressures predicted by Formula (3.2) (from Ref. [6] with permission of Elsevier)

Head no	Dimension			Material			Rupture position	Plastic collapse pressure (MPa)		Difference (%)
	D_i(mm)	D_i/t_m	$D_i/2h_i$	Designation	S_y (MPa)	S_u (MPa)		Experiment, P_{cb}	Formula, P_c	$\frac{P_c - P_{cb}}{P_{cb}} \times 100$
Pressed heads for which rupture occurred in crown										
ZJU-P1	500	88	2	SA738 Gr. B	615	673	Crown	25.5	28.3	11
ZJU-P2	600	97	2	Q245R	332	466	Crown	16.9	17.8	5
ZJU-P3	600	98	2	16MnDR	399	573	Crown	20.2	21.6	7
ZJU-P4	600	98	2	Q345R	414	565	Crown	20.0	21.3	7
ZJU-P5	600	102	2	09MnNiDR	376	500	Crown	18.0	18.1	1
ZJU-P6	1200	203	2	Q245R	332	466	Crown	8.1	8.5	5
ZJU-P7	1200	203	2	Q345R	414	565	Crown	10.3	10.3	0
ZJU-P8	1200	203	2	16MnDR	399	573	Crown	10.1	10.4	3
ZJU-P9	1800	300	2	Q345R	414	565	Crown	6.8	7.0	3
ZJU-P10	1800	310	2	16MnDR	431	575	Crown	6.7	6.9	3
ZJU-P11	1800	310	2	Q245R	332	466	Crown	5.7	5.6	−2
ZJU-P12	1200	600	2	316L2B	335	623	Crown	4.1	3.8	−7
ZJU-P13	600	240	2	S22053	653	803	Crown	13.0	12.4	−5
ZJU-P14	600	214	2	S30403	282	710	Crown	12.7	12.3	−3
ZJU-P15	600	214	2	S30408	269	680	Crown	11.8	11.8	0
ZJU-P16	600	200	2	S32168	230	610	Crown	10.9	11.3	4

(continued)

Table 3.1 (continued)

Head no	Dimension			Material			Rupture position	Plastic collapse pressure (MPa)		Difference (%)
	D_i(mm)	D_i/t_m	$D_i/2h_i$	Designation	S_y (MPa)	S_u (MPa)		Experiment, P_{cb}	Formula, P_c	$\frac{P_c - P_{cb}}{P_{cb}} \times 100$
Pressed heads for which rupture occurred in weld joint										
ZJU-P17	1200	343	2	S30408	382	653	Head-cylinder weld joint	8.0	7.0	−13
ZJU-P18	1800	720	2	S31603	310	562	Head-cylinder weld joint	3.0	2.9	−3
ZJU-P19	600	214	2	S30408	385	689	Head-cylinder weld joint	11.0	11.9	8
ZJU-P20	1800	720	2	S30408	385	689	Head-cylinder weld joint	3.2	3.5	9
ZJU-P21	1200	462	2	S30408	385	689	Head-cylinder weld joint	5.0	5.5	10
ZJU-P22	1800	692	2	S30408	385	689	Head-cylinder weld joint	3.1	3.7	19
ZJU-P23	600	333	2	SUS3042B	402	739	Head-cylinder weld joint	6.7	8.2	22
ZJU-P24	1200	571	2	SUS3042B	444	775	Weld joint in crown	4.1	5.0	22
ZJU-P25	1200	600	2	SUS3042B	444	775	Weld joint in crown	3.6	4.8	33

(continued)

Table 3.1 (continued)

Head no	Dimension			Material			Rupture position	Plastic collapse pressure (MPa)		Difference (%)
	D_i(mm)	D_i/t_m	$D_i/2h_i$	Designation	S_y (MPa)	S_u (MPa)		Experiment, P_{cb}	Formula, P_c	$\frac{P_c - P_{cb}}{P_{cb}} \times 100$
Spun heads										
ZJU-S1	1200	218	2.4	Q345R	405	547	Crown	8.8	9.3	6
ZJU-S2	1200	222	1.7	Q345R	405	547	Crown	8.8	9.1	4
ZJU-S3	1800	545	2.2	S30408	392	708	Head-cylinder weld joint	5.1	4.8	−5
ZJU-S4	1200	480	2.4	S30408	327	700	Head-cylinder weld joint	5.9	5.4	−8
ZJU-S5	1800	529	2.4	S30408	392	708	Head-cylinder weld joint	5.1	5.0	−2
ZJU-S6	1200	462	2.4	S31603	340	605	Head-cylinder weld joint	6.0	4.9	−18
Heads assembled by formed segments										
CBI1	4877	980	2	SA516 Gr. 70	353	552	Weld joint in knuckle	1.6	2.1	31
CBI2	4877	711	2	SA516 Gr.70	361	538	Weld joint in crown	2.3	2.8	22

All tested heads covered a range of five inside diameters, namely 500 mm, 600 mm, 1200 mm, 1800 mm and 4877 mm. The thicknesses of the tested heads were measured by ultrasonic thickness gauges, and the average measured thickness (t_m) of each head was used. The range of diameter-to-thickness ratio (D_i/t_m) is 88–980. Twenty-seven out of the 33 tested heads had radius-to-height ratio ($D_i/2h_i$) of 2, and they were used commonly in engineering application. Radius-to-height ratios of the other tested heads are 1.7, 2.2 and 2.4. The tested heads are constructed by usual manufacturing methods, i.e. cold pressing, cold spinning, and assembly by formed segments. The tested heads covered many kinds of steel including stainless steel, carbon steel and low alloy steel. We obtained the engineering yield strength (S_y) and engineering ultimate tensile strength (S_u) of tensile specimens from the steel plates used to manufacture the tested heads. In addition, the tested heads were produced by different manufacturing companies.

3.2.2 *Rupture Characteristics*

All the tested heads were pressurized until rupture occurred. This section describes the rupture characteristic of the tested heads. Table 3.1 gives the rupture positions, including the head crown (not weld joints), and weld joints in the head crown and head knuckle, and weld joint between the head and the cylinder.

All carbon and low-alloy steel heads fabricated by cold pressing and cold spinning ruptured at the head crown. For example, Head No. ZJU-P1 ruptured with a small crack at the head crown (Fig. 3.3a). For Head No. ZJU-P4, a long crack occurred near the head apex (Fig. 3.3b). Similar rupture occurred at Head No. ZJU-P15 made of stainless steel, as shown in Fig. 3.3c. It can be seen from the long cracks that the heads display obvious ductile fracture.

Most of stainless-steel heads ruptured at weld joints. Figure 3.4a shows that the leak of Head No. ZJU-P20 occurred at the head-cylinder weld joint with a very small crack. The crack of Head No. ZJU-P22 initiated at the head-cylinder weld joint and

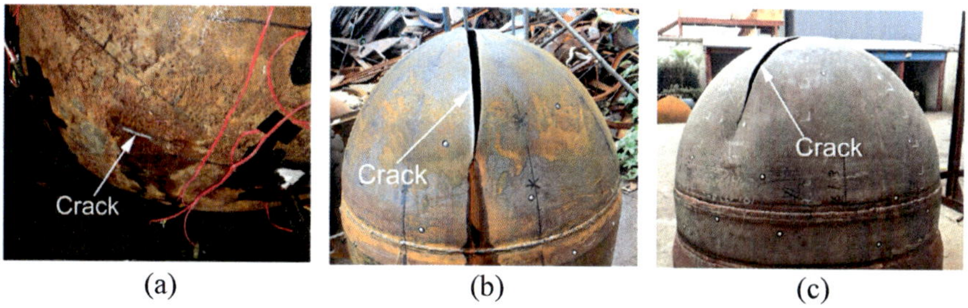

Fig. 3.3 Rupture at head crown: **a** Head No. ZJU-P1; **b** Head No. ZJU-P4; **c** Head No. ZJU-P15 (from Ref. [5] with permission of ASME)

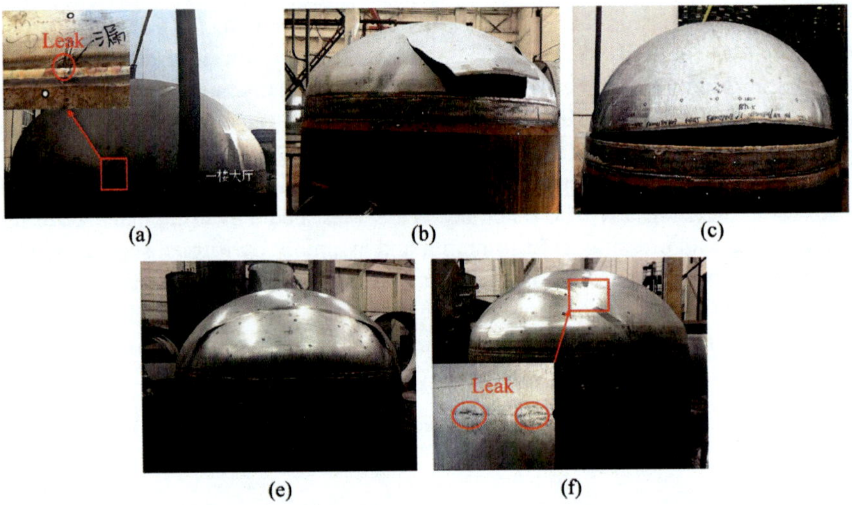

Fig. 3.4 Rupture at weld joints: **a** Head No. ZJU-P20; **b** Head No. ZJU-P22; **c** Head No. ZJU-S6; **d** Head No. ZJU-P24; **e** Head No. ZJU-P25 (from Refs. [6, 7] with permission of Elsevier)

developed along the head (Fig. 3.4b). Head No. ZJU-S6 ruptured along the head-cylinder weld joint with a long crack, as shown in Fig. 3.4c. In addition, the rupture occurred at the weld joint in the head crown for Heads Nos. ZJU-P24 and ZJU-P25, as shown in Fig. 3.4e, f. For the two large-scale heads (Heads Nos. CBI1 and CBI2) tested by Miller et al., the rupture occurred at the weld joints. The rupture of Head No. CBI1 occurred at the weld joints in the knuckle where a buckle formed. Head No. CBI2 ruptured near the head apex when a long crack developed parallel to the weld joint in the crown. It is supposed that the crack of Head CBI2 was located in a heat-affected zone, so the rupture position of Head No. CBI2 is deemed to be the weld joint in the head crown.

3.2.3 *Geometric Strengthening Phenomenon*

Figures 3.5, 3.6, 3.7, 3.8 and 3.9 show the pressure–time curves (loading history) for some typical test heads, as well as their shapes under different pressures. We can see that at lower pressures, the pressure remained constant after pressurization stopped because the test heads were elastic. As the pressure further increased, plastic deformation occurred in the test heads. When pressurization stopped, unlike elastic deformation, plastic deforming did not stop immediately but continued until a stable state was reached, which led to the increase in the volume of the tested head and thus the reduction in pressure because water is almost incompressible under test condition. It can be seen from Figs. 3.5, 3.6, 3.7, 3.8 and 3.9 that the tested ellipsoidal heads

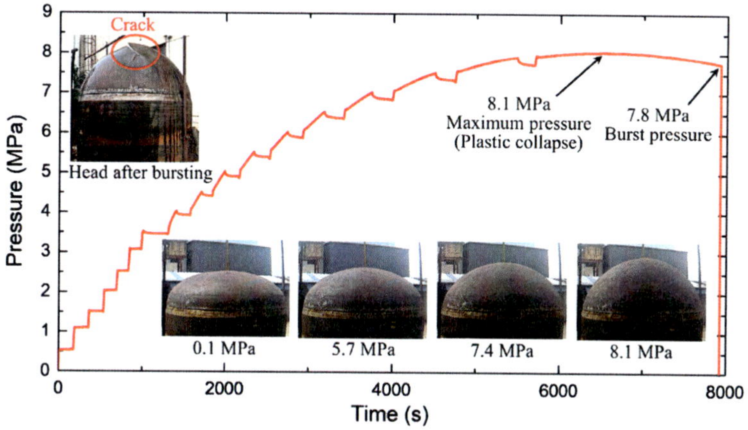

Fig. 3.5 Experimental results of head no. ZJU-P6 (from Ref. [6] with permission of Elsevier)

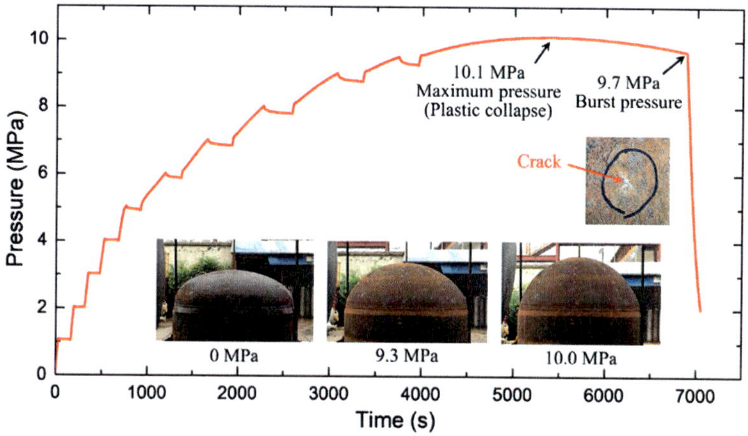

Fig. 3.6 Experimental results of head no. ZJU-P8 (from Ref. [6] with permission of Elsevier)

tended to become hemispherical, indicating that the ellipsoidal heads underwent geometric strengthening.

3.2.4 Plastic Collapse Pressure

The plastic collapse pressure of ellipsoidal heads was determined by the measured pressure–time curves; see Figs. 3.5, 3.6, 3.7, 3.8 and 3.9. Figure 3.5 shows the experimental results of Head No. ZJU-P6. The maximum pressure (8.1 MPa) in the pressure–time curve corresponds to plastic collapse of this tested head, and the

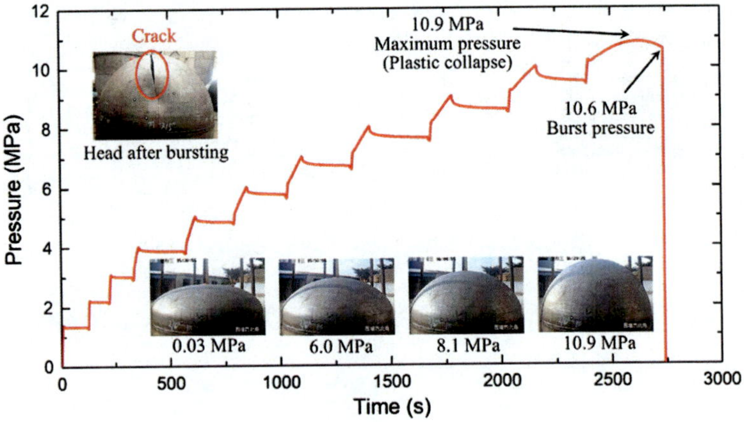

Fig. 3.7 Experimental results of head no. ZJU-P16 (from Ref. [6] with permission of Elsevier)

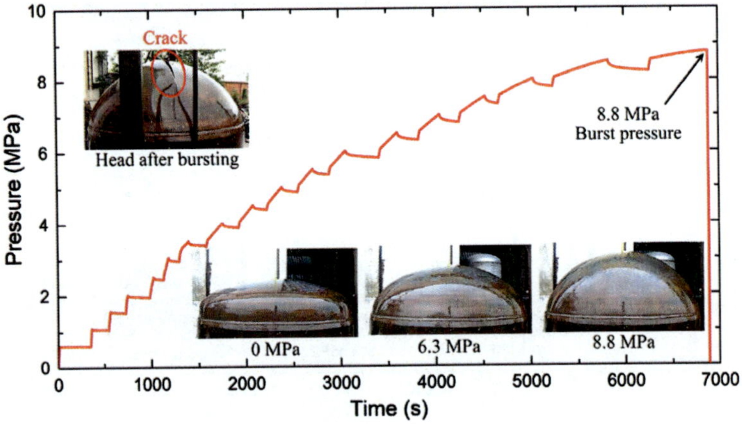

Fig. 3.8 Experimental result of head no. ZJU-S1 (from Ref. [6] with permission of Elsevier)

pressure reduced slightly beyond this maximum point until the head ruptured at pressure of 7.8 MPa, meanwhile a long crack occurred in the head crown. Similar test results were also seen for Head No. ZJU-P8, as shown in Fig. 3.6; the plastic collapse pressure and rupture pressure of Head No. ZJU-P8 were 10.1 MPa and 9.7 MPa respectively. For Head No. ZJU-P16 made of stainless steel, similar results were seen in the test; the plastic collapse pressure and rupture pressure of head No. ZJU-P16 were 10.9 MPa and 10.6 MPa respectively (Fig. 3.7).

For some test heads such as Heads Nos. ZJU-S1 and ZJU-S3, there was no maximum point corresponding to plastic collapse before the heads ruptured, as shown in Figs. 3.8 and 3.9. However, these heads also tended to become hemispherical. In fact, it can be seen from Figs. 3.5, 3.6 and 3.7 that the two types of pressure were

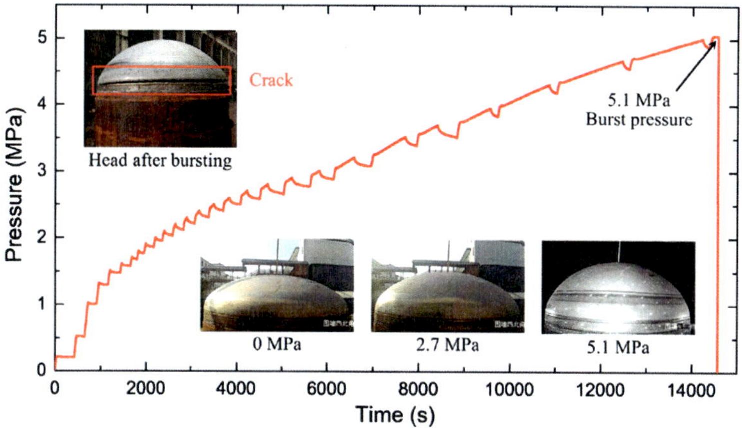

Fig. 3.9 Experimental result of head no. ZJU-S3 (from Ref. [6] with permission of Elsevier)

very close, as mentioned in Sect. 3.1. Thus, the rupture pressure of this kind of head is taken as the plastic collapse pressure. The experimental plastic collapse pressure of Heads No. ZJU-S1 and ZJU-S3 is 8.8 MPa and 5.1 MPa respectively. Table 3.1 lists the experimental plastic collapse pressures of all tested heads.

Deng et al. [9] investigated the effects of loading history on the plastic collapse pressure of thin-walled cylindrical and spherical vessels under internal pressure, and found that loading history does not affect plastic collapse pressure. This conclusion was also verified by uniaxial tensile tests of round bar specimens. Similarly, loading history (Figs. 3.5, 3.6, 3.7, 3.8 and 3.9), including pressure reduction before plastic collapse, does not affect the plastic collapse pressure of test heads.

3.3 Prediction of Plastic Collapse Pressure

3.3.1 Finite Element Model

A two-dimensional (2D) axisymmetric FE model for the tested ellipsoidal head (Head No. ZJU-P1) with a diameter of 500 mm was established to predict the plastic collapse pressure of ellipsoidal heads under internal pressure (Fig. 3.10). The FE model consisted of a tested ellipsoidal head and a thick cylinder, the geometrical parameters of which are shown in Fig. 3.10a. The thick cylinder had such a length that the effect of boundary stresses on the tested head was negligible. Thus, the thick ellipsoidal heads (Fig. 3.2) are not considered in these models. The boundary conditions and applied loading for the FE model are shown in Fig. 3.10a: X direction displacement constraint is imposed on the center of the test head, Y direction displacement constraint is imposed on the bottom edge of the thick cylinder, and increasing

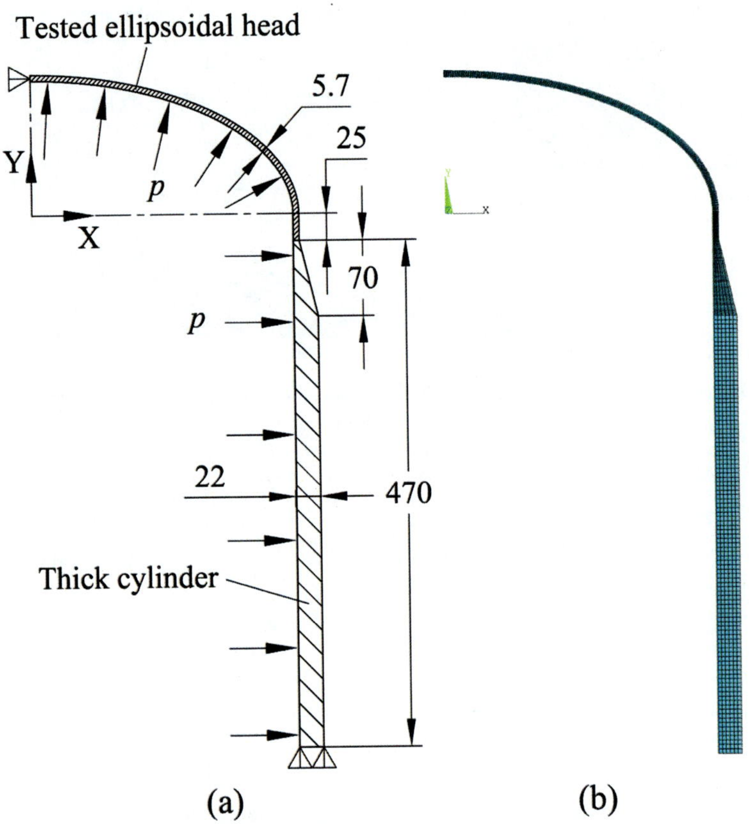

Fig. 3.10 2D axisymmetric FE model for ZJU-P1 head: **a** Geometry and boundary conditions; **b** Mesh model (from Ref. [5] with permission of ASME)

pressure is applied on the inside surface of the model. The element PLANE182 was used to mesh the models, as shown in Fig. 3.10b. PLANE182 is a two-dimensional, four-node element that can be used as an axisymmetric element and has plasticity, large deflection and large strain capabilities.

The tested ellipsoidal head and thick cylinder were made from ASME SA783 Gr. B and Chinese steel Q345R respectively. The true stress-true strain curve of ASME SA783 Gr. B steel was measured from tensile tests (see Fig. 2.4e), while the curve of Q345R was determined by the material model in Annex 3-D of ASME VIII-2 (Fig. 1.3 in Sect. 1.1.2). The effect of strain hardening was considered. The von Mises yield criterion and the flow theory of plasticity were adopted. We used the arc-length method to perform collapse analysis on the FE model, and the effect of geometric nonlinearity was considered in this analysis.

In addition, a three-dimensional (3D) shell FE model for the ZJU-P1 head was also established to predict the plastic collapse pressure of ellipsoidal heads under internal pressure (Fig. 3.11). The geometry and material of 3D shell model were the

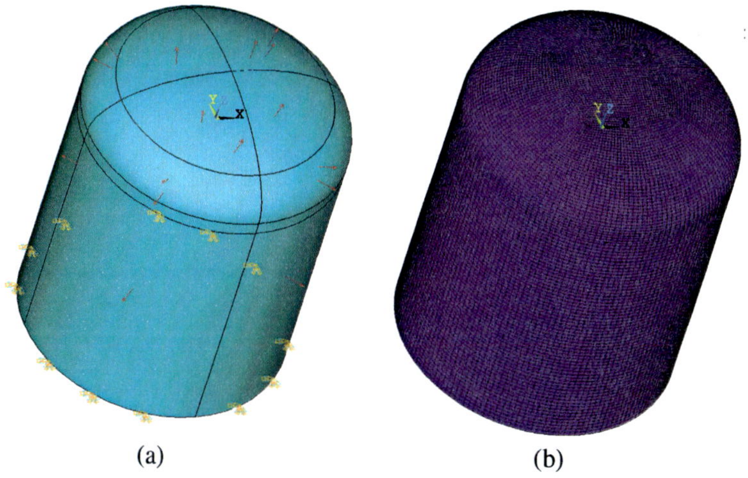

Fig. 3.11 3D shell FE model for the ZJU-P1 head: **a** Geometry and boundary conditions; **b** Mesh model

same as those of the 2D axisymmetric model. As shown in Fig. 3.11a, increasing pressure is imposed on the inside surface of the shell model, and the end of the thick cylinder is fixed. The SHELL181 element is suitable for analyzing thin to moderately thick walled structures and applications of large strain, material nonlinearity, etc. Therefore, the shell FE model was meshed using the SHELL181 element, as shown in Fig. 3.11b.

3.3.2 Finite Element Analysis Results

Figure 3.12 shows the FE results of the plastic collapse of the 2D axisymmetric model for the ZJU-P1 head. It can be seen that as the pressure increased, the head apex deformed significantly, and the ellipsoidal head tended to become hemispherical. When the pressure exceeded the maximum value of 25.1 MPa, it started to decrease, and plastic collapse occurred in the ellipsoidal head. The maximum pressure in the pressure-apex displacement curve was the plastic collapse pressure of the ellipsoidal head; thus, the plastic collapse pressure of the tested ellipsoidal head model was predicted as 25.1 MPa. In addition, the effect of mesh density on the plastic collapse pressure of the 2D axisymmetric FE model was investigated. Figure 3.13 gives the plastic collapse pressures of the 2D axisymmetric FE models with different numbers of elements. It can be seen that the plastic collapse pressure changes little with total number of elements when the number exceeds 660. Therefore, higher mesh density (1980 elements in total) was used for the 2D axisymmetric FE model.

Figure 3.14 shows the FE results of the plastic collapse of the 3D shell model for the tested ellipsoidal head with a diameter of 500 mm. The deformation of the

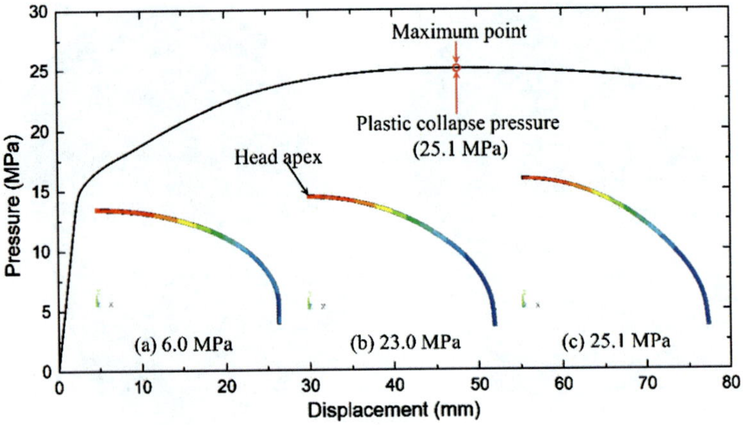

Fig. 3.12 FE results of plastic collapse of 2D axisymmetric model for ZJU-P1 head (from Ref. [5] with permission of ASME)

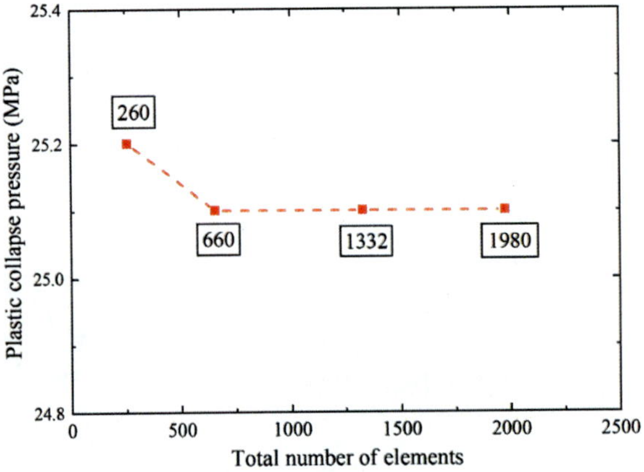

Fig. 3.13 Effect of mesh density on plastic collapse pressure of 2D FE model

3D shell model is similar to that of the 2D axisymmetric model. The maximum pressure in the pressure-apex displacement curve was the plastic collapse pressure of the ellipsoidal head; thus, the plastic collapse pressure of the 3D shell model was predicted as 25.3 MPa. Figure 3.15 gives the plastic collapse pressures of the 3D shell FE models with different numbers of elements. It can be seen that the plastic collapse pressure changes little with total number of elements when the number exceeds 1100. Therefore, higher mesh density was used for the 3D shell FE model.

Table 3.2 presents the comparison between the experimental and predicted plastic collapse pressures of the ZJU-P1 head. The differences between the experimental and predicted plastic collapse pressures are only −1.6% and −0.7% for the 2D axisymmetric model and 3D shell model respectively. This indicates that the FE

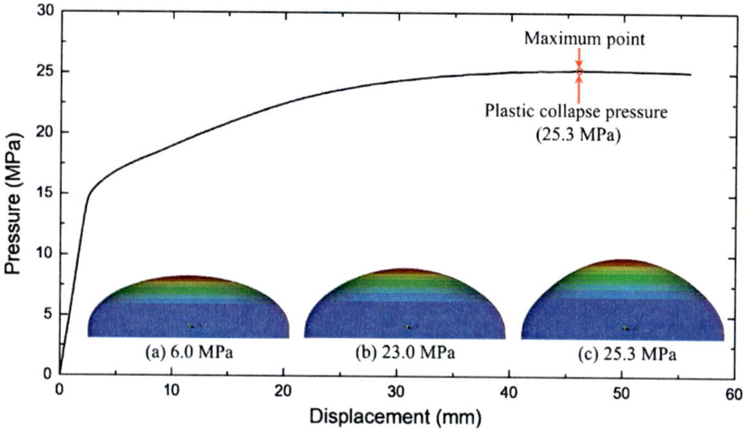

Fig. 3.14 FE result of plastic collapse of 3D shell model for ZJU-P1 head

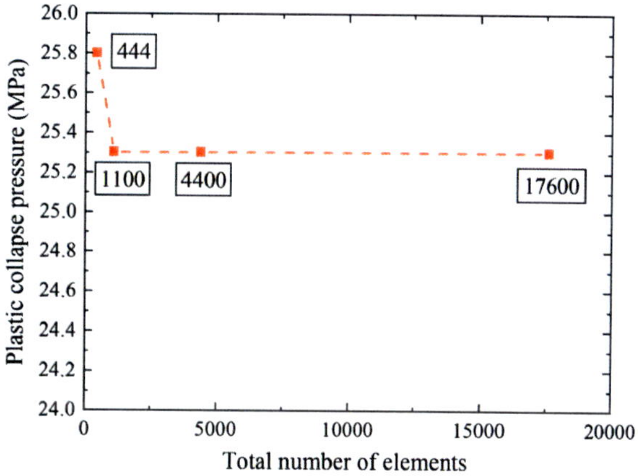

Fig. 3.15 Effect of mesh density on plastic collapse pressure of 3D shell FE model

models, both 2D axisymmetric model or 3D shell model, provide a good prediction of the plastic collapse pressure of ellipsoidal heads. In addition, the FE results in Figs. 3.12 and 3.14 show that the ellipsoidal head tends to become hemispherical, which is similar to the experimental results presented in Sect. 3.2.3.

In addition, although the standard strength of Q345R (per GB/T 150.2) is used in FE models, they provide a good prediction of the plastic collapse pressure of the tested heads (see Table 3.2). We also generated another FE model of the tested head, but its attached thick cylinder was assumed to be made of the same ASME steel used in the tested head. Generally, the measured yield and ultimate tensile strengths of Q345R (not obtained for various reasons) is greater than or equal to the standard strength but does not reach the strength of ASME steel. The predicted plastic collapse pressure of this FE model was 25.6 MPa, slightly higher than that of

Table 3.2 Comparison between experimental and predicted plastic collapse pressures of tested ellipsoidal head (ZJU-P1) with a diameter of 500 mm (from Ref. [5] with permission of ASME)

Experiment, P_{cb} (MPa)	2D axisymmetric model		3D shell model	
	Prediction, P_c (MPa)	Difference (%)	Prediction, P_c (MPa)	Difference (%)
25.5	25.1	−1.6	25.3	−0.7

the FE model of the attached thick cylinder made of Q345R (25.1 MPa). This shows that the properties of the steel used in the attached thick cylinder have little effect on the predicted plastic collapse pressure of the heads. The reason for this is that the attached thick cylinder is thick enough, nearly four times as thick as the head, so it still remains elastic overall even when the head ruptures.

3.3.3 *Influencing Factors of Plastic Collapse Pressure*

3.3.3.1 Effect of Edge Stiffness Due to Attached Component

Ellipsoidal heads are usually attached to two kinds of components in many engineering applications: a cylinder or a bolting flange. If an ellipsoidal head is attached to a bolting flange, we assume that the edge of the head is fixed instead of the bolting flange itself. FE analysis for calculating plastic collapse pressure was performed on ellipsoidal heads with different cylinder lengths. When the cylinder length is 0 mm, it was equivalent to the case of an ellipsoidal head with a flange (fixed head edge). For cases of different diameter-to-thickness ratios ($D_i/t = 20, 60$ and 500), Fig. 3.16 shows the FE results of the plastic collapse pressure of ellipsoidal heads with different

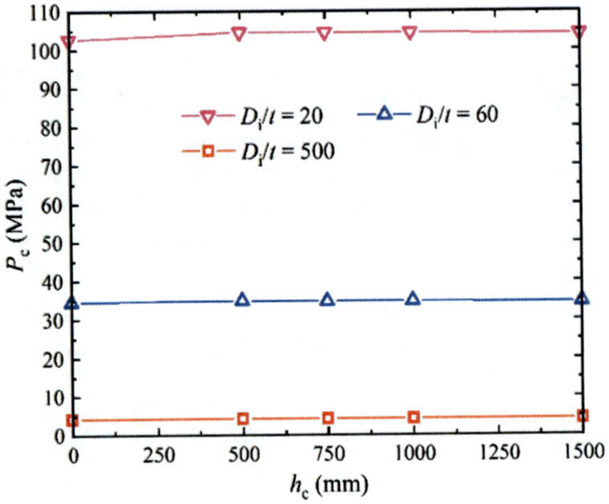

Fig. 3.16 Effects of cylinder length (h_c) on plastic collapse pressure (P_c) (from Ref. [6] with permission of Elsevier)

cylinder lengths. We can see that plastic collapse pressure changes little with cylinder length. Therefore, the effect of edge stiffness due to attached components (cylinder or bolting flange) has little effect on the plastic collapse pressure of ellipsoidal heads.

3.3.3.2 Effects of Geometric Parameters

The geometric structure of an ellipsoidal head is characterized by two geometric parameters: thickness-to-diameter ratio (t/D_i) and radius-to-height ratio ($D_i/2h_i$). To investigate the effects of these two parameters on plastic collapse pressure (P_c), we performed a series of FE analyses on ellipsoidal heads made of various materials for different t/D_i and $D_i/2h_i$. Figure 3.17 shows the effects of thickness-to-diameter ratio and radius-to-height ratio on the plastic collapse pressure of ellipsoidal heads made of all selected materials. We can see that there is a linear relationship between plastic collapse pressure and thickness-to-diameter ratio for various kinds of material. In addition, for each kind of material, the P_c–t/D_i curves are nearly the same for different $D_i/2h_i$, which means that radius-to-height ratio has little effect on the plastic collapse pressure of ellipsoidal heads.

Although radius-to-height ratio has little effect on plastic collapse pressure, we will further discuss on the slight difference. We generated 2D axisymmetric FE models for ellipsoidal heads with different radius-to-height ratios ($D_i/2h_i = 1.4$, 1.7,

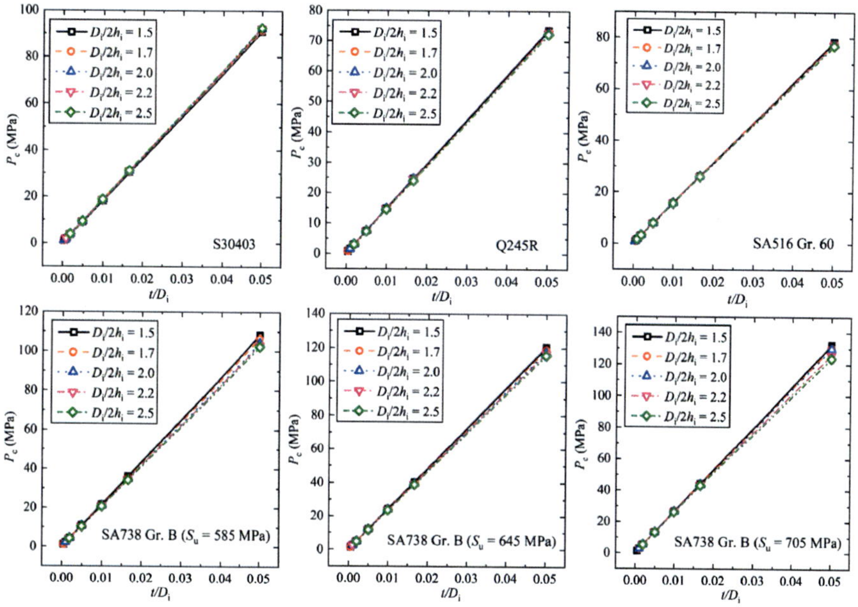

Fig. 3.17 Effect of thickness-to-diameter ratio on plastic collapse pressure of ellipsoidal heads with different materials and different radius-to-height ratios (from Ref. [6] with permission of Elsevier)

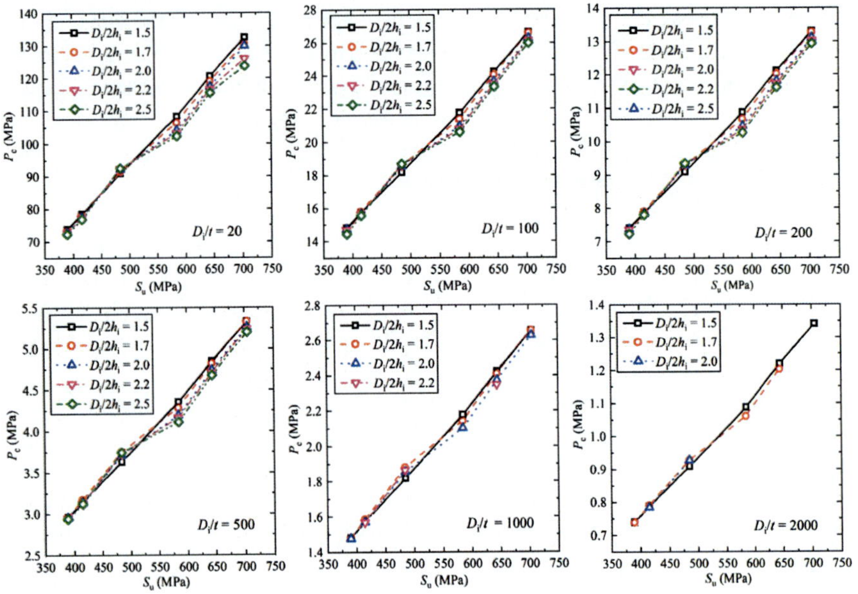

Fig. 3.18 Effects of engineering tensile strength (S_u) on plastic collapse pressure (P_c) (from Ref. [6] with permission of Elsevier)

2.0, 2.2, 2.6 and 3.0). The ellipsoidal heads had the same inside diameter of 500 mm and diameter-to-thickness ratio of 88. Two kinds of steel (S30408 and Q345R) were considered: Q345R is low-alloy steel with a smaller elongation while S30408 is austenitic stainless steel with a larger elongation (see Fig. 1.3). Table 3.3 lists the FE results of ellipsoidal heads with different radius-to-height ratios, including plastic collapse pressure and apex displacement and thickness thinning at plastic collapse pressure. We can see that the plastic collapse pressure shows different trends for various materials, with an increase in radius-to-height ratio. For S30408, with an increase in $D_i/2h_i$, plastic collapse pressure increases; while, for Q345R, with an increase in $D_i/2h_i$, plastic collapse pressure decreases.

As concluded by the experimental and FE results in Sects. 3.2.3 and 3.3.2, ellipsoidal heads tend to become hemisphere while the head wall becomes thinner until plastic collapse occurs. Changing in shape into a hemisphere (shape change) acts as a strengthening factor on plastic collapse pressure, while wall thinning is a weakening factor. The onset of plastic collapse occurs when the following effects are balanced: the strengthening effect due to material hardening and shape change, and the weakening effect due to wall thinning. It can be seen in Table 3.3 that as radius-to-height ratio increased, there was an increase in apex displacement and thickness thinning at plastic collapse pressure. The larger the apex displacement (i.e., the larger the shape change), the more remarkable the strengthening effect, and the larger the thickness thinning (i.e., the smaller the remaining thickness), the more remarkable the weakening effect.

Table 3.3 FE results of plastic collapse of ellipsoidal heads with different materials and radius-to-height ratios (from Ref. [5] with permission of ASME)

Material	Radius-to-height ratio, $D_i/2h_i$	Apex displacement (mm)	Thickness thinning (mm)	Plastic collapse pressure (MPa)
S30408	1.4	94.5	2.12	22.8
	1.7	99.9	2.14	23.0
	2	105.8	2.19	23.2
	2.2	109.3	2.20	23.3
	2.6	114.8	2.22	23.4
	3	120.2	2.23	23.5
Q345R	1.4	51.7	1.37	21.6
	1.7	58.6	1.51	21.2
	2	66.4	1.59	21.0
	2.2	70.2	1.61	20.9
	2.6	77.7	1.66	20.7
	3	84.2	1.72	20.6

For ellipsoidal heads made of S30408 which has a larger elongation, as radius-to-height ratio increased, thickness thinning had a much smaller increase of 5%, but apex displacement was much larger, compared with heads made of Q345R; that is, shape change has a much more remarkable effect than thickness thinning. When radius-to-height ratio is larger, apex displacement is larger (i.e., shape change is larger), and consequently the strengthening effect of shape change on plastic collapse pressure is more remarkable, which leads to higher plastic collapse pressure. Therefore, for ellipsoidal heads made of S30408, plastic collapse pressure increases with an increase in radius-to-height ratio.

For ellipsoidal heads made of Q345R which has a smaller elongation, as radius-to-height ratio increased, there was a larger increase of 35% in thickness thinning, but apex displacement was smaller, compared with heads made of S30408; that is, thickness thinning has much more remarkable effect than shape change. When radius-to-height ratio is larger (i.e., head height is smaller), thickness thinning is larger (i.e., remaining thickness is smaller), and consequently the weakening effect of thickness thinning on plastic collapse pressure is more remarkable, which leads to lower plastic collapse pressure. Therefore, for Q345R, plastic collapse pressure decreases with an increase in radius-to-height ratio.

3.3.3.3 Effects of Material Parameters

It is generally known that engineering tensile strength significantly affects plastic collapse pressure. As shown in Fig. 3.18, engineering tensile strength increases the plastic collapse pressure of ellipsoidal heads, and there is an approximately linear relationship between engineering tensile strength and plastic collapse pressure for different geometrical parameters (thickness-to-diameter ratio t/D_i and radius-to-height ratio $D_i/2h_i$).

3.4 Formula for Predicting Plastic Collapse Pressure

3.4.1 Parameter Study

Based on the FE models presented in Sect. 3.3.1, a further parametric study was performed to obtain a formula for predicting the plastic collapse pressure of ellipsoidal heads under internal pressure. The scope of geometric parameters for ellipsoidal heads was determined with reference to pressure vessel codes and standards such as EN 13345-3 and ASME VIII-2. In EN 13345-3, the ranges of applicability of the design method for plastic collapse of ellipsoidal heads are $1.7 < D_i/2h_i < 2.2$, and $12.5 \leq D_o/t \leq 1000$, where D_o is the outer diameter of the heads. In ASME VIII-2, the ranges are $1.7 \leq D_i/2h_i \leq 2.2$, $20 \leq L/t \leq 2000$, and $0.7 \leq L/D_i \leq 1.0$, where L is the crown radius of the torispherical heads. Ellipsoidal heads with a smaller or larger diameter-to-thickness ratio ($D_i/t < 20$ or $D_i/t > 2000$) are barely used in practice. Therefore, the ranges of the geometric parameters in this study were determined as follows: $1.5 \leq D_i/2h_i \leq 2.5$, and $20 \leq D_i/t \leq 2000$ ($0.0005 \leq t/D_i \leq 0.05$).

As shown in Table 3.4, we selected several kinds of usual material for steel ellipsoidal heads. S30403 is a stainless steel plate for pressure vessels and general applications. Q245R is Chinese carbon steel commonly used in pressure vessel components. SA516 Gr. 60 is an ASME carbon steel for moderate- and low-temperature service. SA738 Gr. B is an ASME carbon-manganese-silicon steel with a high strength; for example, it is used to construct large-scale ellipsoidal heads for nuclear steel containment [10]. The true stress-true strain curves of these steels were determined using the

Table 3.4 Material parameters used in FE models to calculate plastic collapse pressure (from Ref. [6] with permission of Elsevier)

Material	Young's modulus E (GPa)	Engineering yield strength S_y (MPa)	Engineering tensile strength S_u (MPa)
S30403	195	170	485
Q245R	201	205	390
SA516 Gr. 60	202	220	415
SA738 Gr. B	200	415	585
	200	415	645
	200	415	705

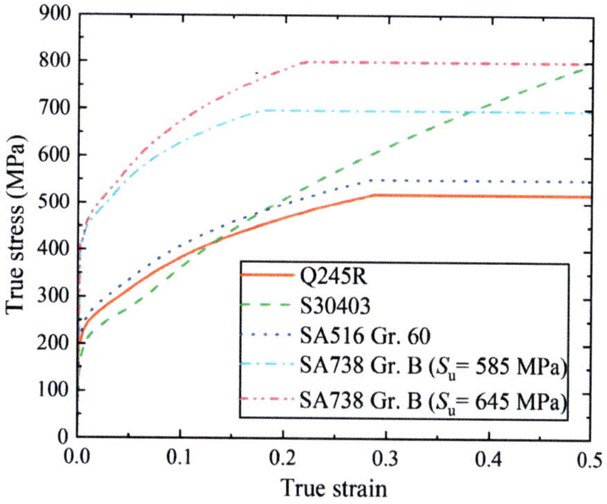

Fig. 3.19 True stress-true strain curves of various materials used in FE models to calculate plastic collapse pressure (from Ref. [6] with permission of Elsevier)

material model in ASME VIII-2 (Sect. 1.1.2). Figure 3.19 shows the true stress-true strain curves of these steels which were used in the FE models to calculate the plastic collapse pressure of ellipsoidal heads. The true stress-true strain curves were determined by modulus of elasticity, engineering yield strength and engineering tensile strength which were obtained from the material properties tables in ASME II-D and GB/T 150.2. Table 3.4 lists the material properties of these steels. For SA7380 Gr. B, different values of engineering tensile strength were considered in the range specified in ASME II-A [11]. The curve of SA7380 Gr. B with $S_u = 705$ MPa is shown in Fig. 2.6a.

We carried out finite element analysis on the calculation of plastic collapse pressure for a total of 185 ellipsoidal head cases within the scope presented in Sect. 3.4.1. As concluded in Sect. 3.3.3, engineering tensile strength (S_u) and thickness-to-diameter ratio (t/D_i) significantly affect plastic collapse pressure (P_c). Thus, the two parameters (S_u and t/D_i) need to be considered in the development of the formula for predicting the plastic collapse pressure of ellipsoidal heads. As shown in Fig. 3.20, dimensionless parameter (P_c/S_u) has a linear relationship with t/D_i for cases of various materials.

3.4.2 Development of New Formula Considering Nonlinearity Strengthening

Based on the above parameter study, we used a Formula (3.1) to express the linear relationship between P_c/S_u and t/D_i. Coefficient C of 3.7 was determined by linear curve fitting; see Fig. 3.21. The coefficient of determination (adjusted R-Square) was used to evaluate the goodness of fitting. The closer the coefficient of determination is the value of 1, the closer the fitting is the data points. For the fitting of

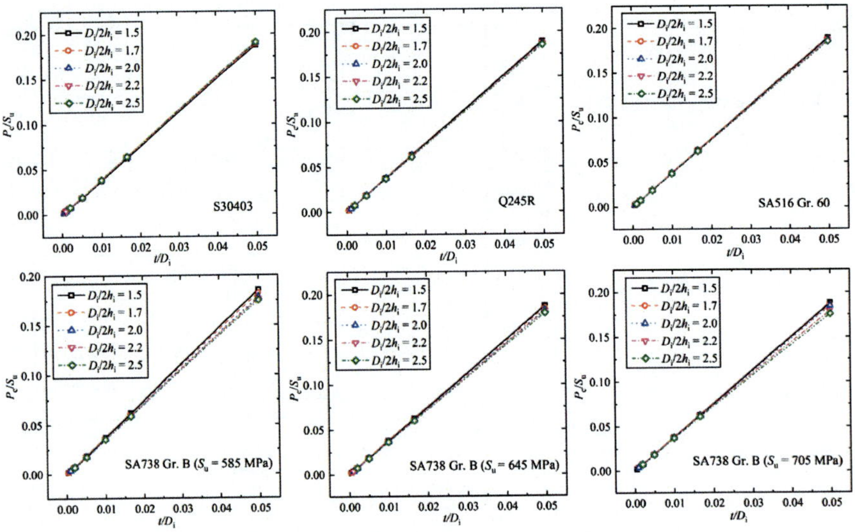

Fig. 3.20 Relationship between P_c/S_u and t/D_i (from Ref. [6] with permission of Elsevier)

Fig. 3.21 Curve fitting for FE results of plastic collapse pressure of ellipsoidal heads (from Ref. [6] with permission of Elsevier)

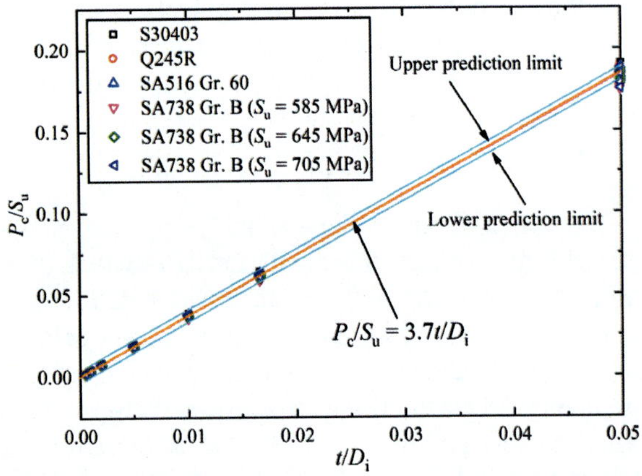

Formula (3.2), the coefficient of determination (adjusted R-Square) is 0.999, which means that this fitting is good. The 95% prediction bands for the numerical results is also shown in Fig. 3.21. Therefore, we proposed a new formula for predicting the plastic collapse pressure of ellipsoidal heads under internal pressure, which takes nonlinearity strengthening into account; see Formula (3.2). This formula is limited as follows: $1.5 \leq D_i/2h_i \leq 2.5$ and $20 \leq D_i/t \leq 2000$. This formula is also applicable to ellipsoidal heads made of steel including stainless steel, carbon steel and low alloy steel specified in pressure vessel codes and standards.

$$\frac{P_c}{S_u} = C\frac{t}{D_i} \tag{3.1}$$

$$P_c = 3.7\frac{t}{D_i}S_u \tag{3.2}$$

3.4.3 Experiment Verification

A comparison of the plastic collapse pressures predicted by Formula (3.2) and the experimental results is shown in Table 3.1 and Fig. 3.22. It can be seen that the difference between experimental (P_{cb}) and predicted (P_c) plastic collapse pressures varies from −18 to 33%, and the average difference is 5%. The difference is lower than ±15% for 26 out of the 33 tested heads. Therefore, the new Formula (3.2) proposed in this paper provides an accurate prediction of the plastic collapse pressure of ellipsoidal heads under internal pressure.

The new Formula (3.2) provides a more accurate prediction for the tested heads which ruptured at the crown but not at the weld joints because these heads experience enough deformation to become an approximate hemisphere before rupturing. In addition, some test heads (for example, heads Nos. ZJU-P23, ZJU-P24, ZJU-P25, CBI1 and CBI2) ruptured prematurely at a much lower pressure than the predicted plastic collapse pressure, and they ruptured at weld joints. One cause may be poor

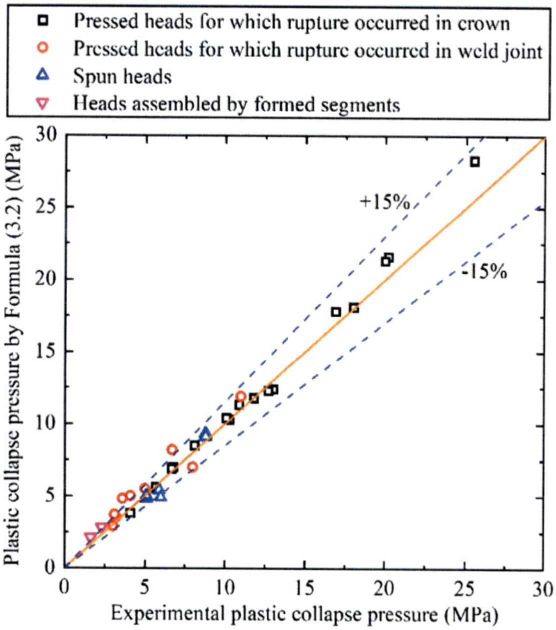

Fig. 3.22 Comparison of plastic collapse pressures predicted by new Formula (3.2) and experimental results (from Ref. [6] with permission of Elsevier)

weld joints which do not meet the welding requirement of pressure vessel codes and standards that the tensile strength of weld joint should equal or exceed that of the base metals to be joined.

3.4.4 *Analysis and Discussion*

3.4.4.1 Effect of Material Strain Hardening

As mentioned in Sect. 1.3.1, the effect of material strain hardening is not considered in the prediction of plastic collapse pressure of ellipsoidal heads under internal pressure. In order to discuss the effect of material strain hardening, for case of the tested ellipsoidal head (ZJU-P1) with a diameter of 500 mm, four types of FE analyses were performed using 2D axisymmetric model presented in Sect. 3.3.1: elastic analysis, limit analysis, perfectly plastic analysis and plastic collapse analysis.

For elastic analysis, the head failed at pressure of 14.1 MPa when the circumferential stress at the middle surface of the head apex reached the engineering yield strength of 615 MPa. Limit analysis without geometric nonlinearity was performed using perfectly plastic material; thus, the limit pressure was determined as 15.3 MPa by the pressure-apex displacement curve (Fig. 3.23) obtained from limit analysis. Perfectly plastic analysis was performed using perfectly plastic material, and the effect of geometric nonlinearity was considered in this analysis. Figure 3.24 shows the pressure-apex displacement curve obtained from perfectly plastic analysis. It can be seen in Fig. 3.24 that the plastic pressure was 16.7 MPa by two elastic

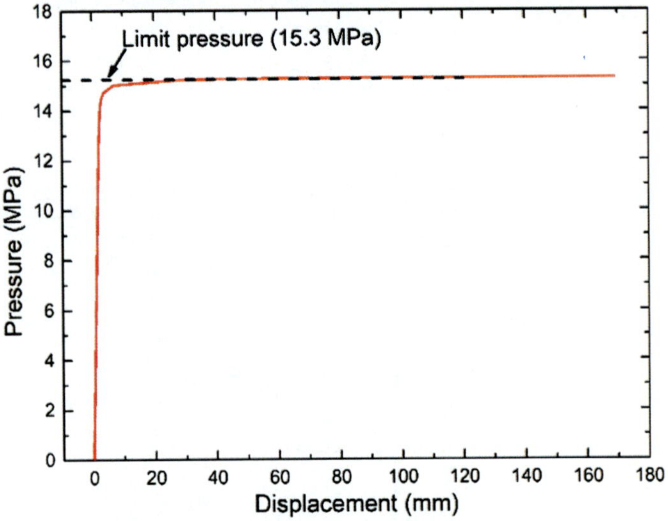

Fig. 3.23 Pressure versus apex displacement curve from limit analysis

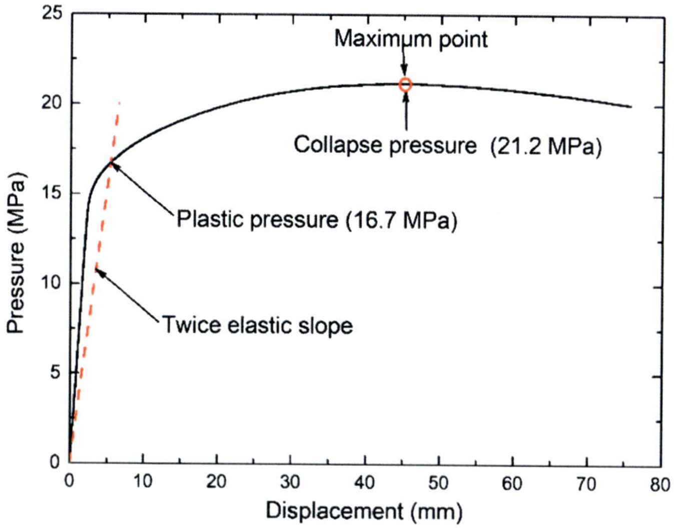

Fig. 3.24 Pressure versus apex displacement curve from perfectly plastic analysis

slope method, and the collapse pressure corresponding to the maximum point was 21.2 MPa. Sections 3.3.1 and 3.3.2 present plastic collapse analysis for this ellipsoidal head, considering the effects of material hardening and geometric nonlinearity; and the plastic collapse pressure was calculated as 25.1 MPa.

Figure 3.25 shows a comparison of failure pressures determined by different analysis methods and test. It is clear that the failure pressures determined by elastic analysis, limit analysis and perfectly plastic analysis are lower than experimental

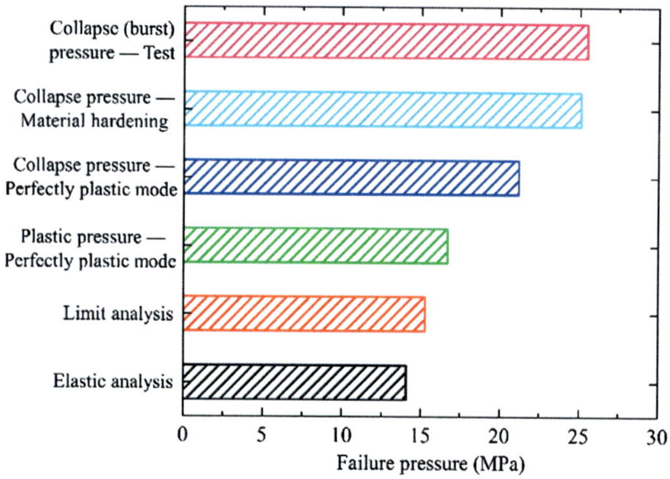

Fig. 3.25 Comparison of failure pressures determined by different analysis methods and test data

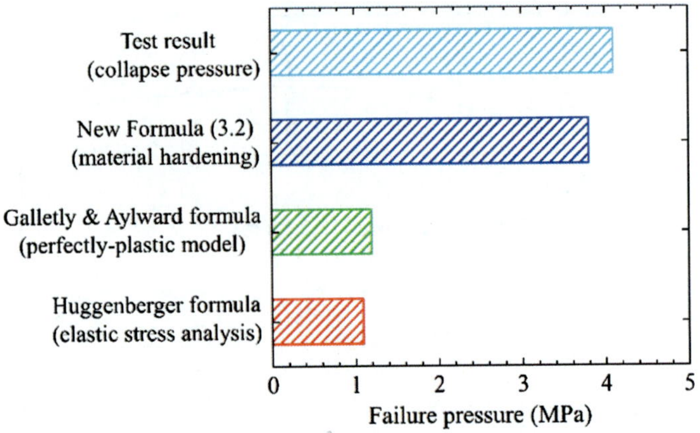

Fig. 3.26 Comparison of failure pressures predicted by different formulas and test results

pressure because the material hardening effect was not considered in these analyses. And the plastic collapse analysis, considering material hardening effect, provides a better prediction of the plastic collapse pressure of ellipsoidal heads under internal pressure.

Furthermore, the material strain hardening effect is not considered in the existing formulas for predicting plastic collapse, as presented in Sect. 1.3.2. The Huggenberger Formula (1.13) is based on elastic stress analysis. The Galletly and Aylward Formula (1.17) is based on plastic pressure numerical results using perfectly plastic material, and this formula is only applicable to 2:1 ellipsoidal head in the range of $500 < D_i/t < 1500$ and $207\ \text{MPa} < S_y < 414\ \text{MPa}$. Figure 3.26 shows a comparison of failure pressures predicted by different formulas and test result for the tested ellipsoidal head (for example, Head No. ZJU-P12 with D_i/t of 600 and S_y of 335 MPa). It is clear that the failure pressures determined by the Huggenberger and Galletly and Aylward formulas are lower than experimental data because material hardening effect was not considered in these formulas. The new Formula (3.2), considering material hardening effect, provides a better prediction of the plastic collapse pressure of ellipsoidal heads under internal pressure. In addition, Formulas (1.15) and (1.16) for predicting plastic collapse pressure based on the stress ratio (K) are used in codes and standards, so the comparison of plastic collapse pressures between these formulas and experimental data will be discussed in Sect. 4.1.3.

3.4.4.2 Comparison of Plastic Collapse Between Ellipsoidal and Equivalent Torispherical Heads

In pressure vessel codes and standards such as ASME VIII-2 and EN 13445-3, ellipsoidal heads are designed as equivalent torispherical heads using geometric equivalency approaches, as mentioned in Sect. 1.4.1. 2D axisymmetric FE models

Table 3.5 Comparison of plastic collapse pressures between ellipsoidal and equivalent torispherical heads (from Ref. [5] with permission of ASME)

D_i/t	$D_i/2h_i$	Plastic collapse pressure (MPa)		Difference (%)
		Ellipsoidal head	Equivalent torispherical head	
S30408				
88	1.4	22.8	21.6	5.6
	1.7	23.0	21.9	5.0
	2	23.2	22.2	4.5
	2.2	23.3	22.2	5.0
	2.6	23.4	22.3	4.9
	3.0	23.5	22.4	4.9
200	1.4	9.92	9.68	2.5
	1.7	10.0	9.81	1.9
	2	10.1	9.89	2.1
	2.2	10.1	9.91	1.9
	2.6	10.1	9.93	1.7
	3.0	10.1	9.92	1.8
500	1.4	3.98	3.91	1.8
	1.7	4.02	3.95	1.8
	2	4.04	3.98	1.5
	2.2	4.04	3.99	1.3
	2.6	4.04	3.99	1.3
	3.0	4.04	3.98	1.5
Q345R				
88	1.4	21.6	21.1	2.4
	1.7	21.2	20.8	1.9
	2	21.0	20.5	2.4
	2.2	20.9	20.4	2.5
	2.6	20.7	20.2	2.5
	3.0	20.6	20.1	2.5
200	1.4	9.49	9.37	1.3
	1.7	9.30	9.20	1.1
	2	9.17	9.06	1.2
	2.2	9.11	9.0	1.2
	2.6	9.02	8.91	1.2
	3.0	8.96	8.85	1.2
500	1.4	3.81	3.76	1.3
	1.7	3.73	3.69	1.1

(continued)

Table 3.5 (continued)

D_i/t	$D_i/2h_i$	Plastic collapse pressure (MPa)		Difference (%)
		Ellipsoidal head	Equivalent torispherical head	
	2	3.68	3.63	1.4
	2.2	3.65	3.61	1.1
	2.6	3.62	3.57	1.4
	3.0	3.59	3.55	1.1
SA738 Gr. B				
88	1.4	27.5	27.1	1.5
	1.7	26.1	25.6	2.0
	2	25.4	24.9	2.0
	2.2	25.2	24.6	2.4
	2.6	25.0	24.4	2.5
	3.0	24.9	24.2	2.9
200	1.4	12.1	11.9	1.7
	1.7	11.4	11.2	1.8
	2	11.1	10.9	1.8
	2.2	11.0	10.8	1.9
	2.6	10.8	10.7	0.9
	3.0	10.8	10.6	1.9
500	1.4	4.85	4.78	1.5
	1.7	4.57	4.50	1.6
	2	4.44	4.38	1.4
	2.2	4.40	4.34	1.4
	2.6	4.34	4.28	1.4
	3.0	4.31	4.25	1.4

for ellipsoidal and their equivalent torispherical heads were established to compare plastic collapse pressures of two types of the heads. The heads are assumed to have the same diameter of 500 mm. We considered three diameter-to-thickness ratios ($D_i/t = 88$, 200 and 500), six radius-to-height ratios ($D_i/2h_i = 1.4$, 1.7, 2.0, 2.2, 2.6 and 3.0), and the three material models (S30408, Q345R and SA738 Gr. B). The calculated plastic collapse pressures are compared in Table 3.5, and it is shown that the differences are no more than 6%, meaning that there is little difference of plastic collapse pressures between ellipsoidal and their equivalent torispherical heads subjected to internal pressure. The reason for this is that the overall deformation is similar for two types of heads: The apex displacement of equivalent torispherical head is close to that of ellipsoidal head, especially when pressure is higher, as shown in Fig. 3.27. Furthermore, the stress state is similar for the two types of heads when pressure is higher (Fig. 1.8b).

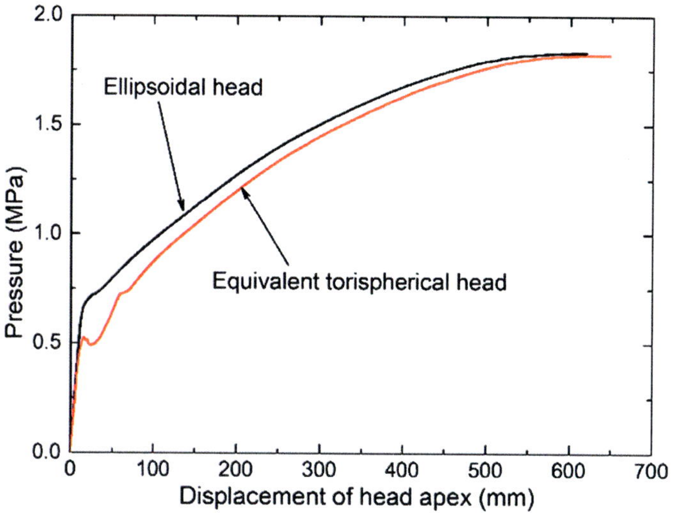

Fig. 3.27 Pressure-apex displacement curves of ellipsoidal and equivalent torispherical heads (from Ref. [5] with permission of ASME)

The two approximately ellipsoidal heads (Heads Nos. CBI1 and CBI2) tested by Miller et al. were actual torispherical heads with crown and knuckle radiuses equal to the values determined by geometric equivalency approaches when the radius-to-height ratio is 2. Since there is little difference in plastic collapse pressure between torispherical and ellipsoidal heads satisfying the geometric equivalencies, the two heads tested by Miller et al. were used as test data to investigate plastic collapse of ellipsoidal heads.

In addition, since there is little difference in plastic collapse pressure between ellipsoidal and equivalent torispherical heads, the new Formula (3.2) is applicable to torispherical heads satisfying the geometric equivalencies (see Sect. 1.4.1).

References

1. Updike DP, Kalnins A (1994) Burst by tensile instability of vessels with torispherical head. ASME PVP Conf 277:89–94
2. Updike DP, Kalnins A (1998) Tensile plastic instability of axisymmetric pressure vessels. ASME J Press Vessel Technol 120(1):6–11
3. Liu P, Zheng J, Ma L, Miao C, Wu L (2008) Calculations of plastic collapse load of pressure vessel using FEA. J Zhejiang Univ Sci A 9(7):900–906
4. Kadam M, Gopalsamy B, Bujurke AA, Joshi KM (2018) Estimation of static burst pressure in unflawed high pressure cylinders using nonlinear FEA. Thin-Walled Struct 126:79–84
5. Zheng J, Yu Y, Chen Y, Li K, Zhang Z, Peng W, Gu C, Xu P (2021) Comparison of ellipsoidal and equivalent torispherical heads under internal pressure: buckling, plastic collapse and design rules. ASME J Press Vessel Technol 143:021301

6. Zheng J, Zhang Z, Li K, Yu Y, Gu C (2020) A simple formula for prediction of plastic collapse pressure of steel ellipsoidal heads under internal pressure. Thin-Walled Struct 156:106994
7. Zhang Z (2021) Research on failure mechanism-based method for design of steel ellipsoidal heads under internal pressure. PhD thesis, Zhejiang University, Hangzhou
8. Miller CD, Grove RB, Bennett JG (1986) Pressure testing of large-scale torispherical heads subjected to knuckle buckling. Int J Press Vessel Pip 22:147–159
9. Deng Y, Chen G (2011) Carrying capacity of strain-hardening austenitic stainless steel pressure vessels under hydrostatic pressure. ASME PVP conference, paper no PVP2011–57216
10. Westinghouse Electric Co, LLC (2011) Westinghouse AP1000 design control document (rev 19) US Nuclear Regulatory Commission, Washington, DC, 2011, Accession No ML11171A500
11. Committee ANS (2019) ASME BPVC II-A: ASME boiler and pressure vessel code, Section II, Part A, ferrous material specifications. The American Society of Mechanical Engineering, New York

Chapter 4
New Method for Design of Ellipsoidal Heads

4.1 Problems of Current Design Methods

4.1.1 *Local Buckling Criteria*

Local buckling criteria are intended to determine whether internally pressurized ellipsoidal heads will buckle or not. Table 4.1 shows a summary of local buckling criteria. As presented in Sect. 2.4.4, we developed a new buckling criterion on the basis of FE analysis and experimental data. Buckling is considered when the ratios of diameter-to-thickness and radius-to-height are located at the orange curve and/or the top-right corner, according to the new criterion shown in Fig. 4.1 (Fig. 2.51 in Sect. 2.4.4). As stated in Sect. 1.4, some pressure vessel codes and standards provide buckling criteria for ellipsoidal heads. In EN 13445–3, for heads in the range of $1.7 < D_i/2h_i < 2.2$, local buckling is considered in head design when $D_i/t > 200$. In ASME VIII-1, local buckling is considered in head design when $D_i/t > 500$. In Chinese standard GB/T 150.3, for $D_i/2h_i \leq 2.0$, local buckling is considered in head design when $D_i/t > 666$; while for $D_i/2h_i > 2.0$, local buckling is considered when $D_i/t > 333$. Buckling is considered when $D_i/2h_i > 1.414$ in accordance with the Huggenberger formula (see Formula (1.9) in Sect. 1.2.1).

Figure 4.1 shows a comparison of different local buckling criteria and test data. We can see that EN 13445–3 provides a more conservative and limited buckling criterion; some heads do not buckle when $D_i/t > 200$, and there is no comment on this criterion when $D_i/2h_i \leq 1.7$ or $D_i/2h_i \geq 2.2$. The buckling criterion in ASME VIII-1 is inaccurate because buckling occurs in heads with higher $D_i/2h_i$ even if $D_i/t \leq 500$; for example, buckling occurred in the Heads Nos. ZJU-S6 and ZJU-S4 with $D_i/2h_i = 2.4$ and $D_i/t = 400$. The buckling criteria in GB/T 150.3 is also inaccurate because buckling occurs in ellipsoidal heads with $D_i/2h_i > 2.2$ and $D_i/t \leq 333$; for example, the Heads Nos. RA2 and RA4 with $D_i/2h_i = 2.5$ and $D_i/t = 250$. In addition, the buckling criterion determined by the Huggenberger formula is also more conservative and limited; No buckling occurs in some heads such as Heads Nos. ZJU-FN1 and ZJU-FN2 with $D_i/2h_i = 1.6$, even if $D_i/2h_i > 1.414$,

J. Zheng and K. Li, *New Theory and Design of Ellipsoidal Heads for Pressure Vessels*,
https://doi.org/10.1007/978-981-16-0467-6_4

Table 4.1 Summary of buckling criteria (from Ref. [1])

Name	Buckling criteria
New criterion (Zheng & Li)	Figure 4.1 (Fig. 2.51 in Sect. 2.4.4)
EN 13445–3	$D_i/t > 200$, $1.7 < D_i/2h_i < 2.2$
ASME VIII-1	$D_i/t > 500$
GB/T 150.3	$D_i/2h_i \le 2.0$, $D_i/t \le 666$; $D_i/2h_i > 2.0$, $D_i/t \le 333$
Huggenberger formula	$D_i/2h_i > 1.414$

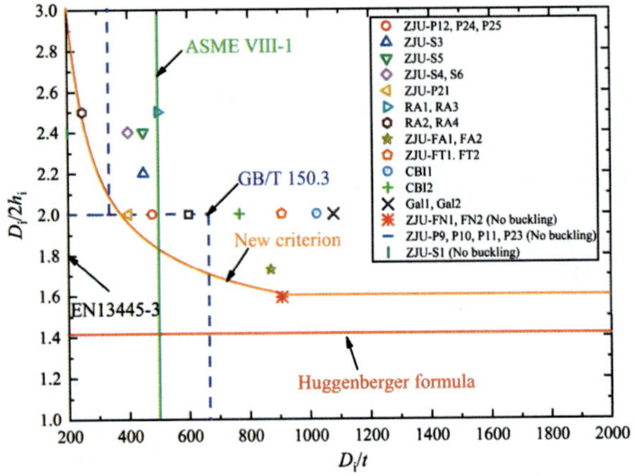

Fig. 4.1 Comparison of different local buckling criteria and test data (from Ref. [1])

and the effect of the diameter-to-thickness ratio is not considered. In summary, there are problems of the local buckling criteria which are more conservative, limited, and/or inaccurate in the current design codes and standards including EN 13445–3, ASME VIII-1 and GB/T 150.3, as well as Huggenberger formula. The comparison in Fig. 4.1 shows that the new local buckling criterion is more accurate and reasonable than the others, making it more applicable to determine whether internally pressurized ellipsoidal heads will buckle or not.

Section 4.1.1 is from the paper [1].

4.1.2 Geometric Equivalence as Torispherical Heads

In ASME VIII-1, ellipsoidal heads with $t/L < 0.002$ are designed as equivalent torispherical heads to prevent the failure of the knuckle. The equivalent approach used in ASME VIII-1 is presented in Table 1.2. In ASME VIII-2, ellipsoidal heads with $1.7 \le D_i/2h_i \le 2.2$ are designed as equivalent torispherical heads using equivalent Formulas (1.25) and (1.26). EN13445-3 also uses equivalent Formulas (1.25) and

(1.26) for the geometric equivalency between ellipsoidal and torispherical heads subjected to internal pressure. As concluded in Sect. 3.4.4.2, there is little difference of plastic collapse pressures between ellipsoidal and equivalent torispherical heads. Comparison of buckling pressures between ellipsoidal and equivalent torispherical heads is presented below.

To compare buckling pressures between an ellipsoidal head and its corresponding equivalent torispherical head based on the above geometric equivalency, a series of FE head models with perfect shape were generated to compute the buckling pressures of the two types of heads. The FE head models cover three diameter-to-thickness ratios ($D_i/t = 500$, 1000 and 1500) and five radius-to-height ratios ($D_i/2h_i = 1.7, 2.0, 2.2,$ 2.6 and 3.0), and three types of material (S30408, Q345R and SA738 Gr. B) used in the FE head models are considered. The stress–strain curves are shown in Fig. 1.3 for S30408 and Q345R in Fig. 2.4e for SA738 Gr. B. The calculated buckling pressures of the ellipsoidal and corresponding equivalent torispherical heads are shown in Table 4.2. We can see from Table 4.2 that the buckling pressures of ellipsoidal heads are more than 17% greater than those of the corresponding equivalent torispherical heads. It can also be seen that when the radius-to-height ratio is relatively low, no buckling occurs in the ellipsoidal heads, but buckles form in the equivalent torispherical heads. This can be explained by the fact that the peak value and range of the compressive stresses in the ellipsoidal heads are smaller than those of the equivalent torispherical heads, as discussed in Sect. 1.2.3. Therefore, the ellipsoidal heads have more buckling resistance than the equivalent torispherical heads which are determined by the geometric equivalent approaches in pressure vessel codes and standards; similar result is given by Galletly [2]. Although the current design rules for buckling failure based on the geometric equivalency approaches are on the conservative side, this results in uneconomical design. Therefore, it is recommended that a design equation for protecting against buckling should be proposed only for ellipsoidal heads subjected to internal pressure, rather than using the geometric equivalent approaches.

Section 4.1.2 is from Ref. [3] with permission of ASME.

4.1.3 Strengthening Effect of Nonlinearity on Strength

The strength of an ellipsoidal heads constitutes its ability to prevent plastic collapse. Section 1.4.2 reviews design formulas for protecting against the plastic collapse of ellipsoidal heads in the codes and standards ASME VIII-1 and VIII-2, EN 13445–3, GB/T 150.3 and JB 4732. Failure pressures corresponding to plastic collapse for the codes and standards are determined using the design formulas by substituting allowable stress for yield strength or ultimate tensile strength. With respect to the strength of head crowns, ASME VIII-2 and EN 13445–3 use the same design method which is based on the prediction formula in which the head fails when the primary membrane stress in the head crown reaches the ultimate tensile strength of material; thus, the failure pressures corresponding to the strength of the crown for ASME

Table 4.2 Comparison of buckling pressures between ellipsoidal and equivalent torispherical heads (from Ref. [3] with permission of ASME)

D_i/t	$D_i/2h_i$	Buckling pressure (MPa)		Difference (%)
		Ellipsoidal head	Torispherical head	
S30408				
500	1.7	NB	NB	–
	2	NB	NB	–
	2.2	NB	NB	–
	2.6	NB	0.83	–
	3.0	NB	0.73	–
1000	1.7	NB	NB	–
	2	0.68	0.47	44.7
	2.2	0.50	0.36	38.9
	2.6	0.35	0.25	40.0
	3.0	0.28	0.19	47.4
1500	1.7	NB	NB	–
	2	0.32	0.23	39.1
	2.2	0.26	0.17	52.9
	2.6	0.18	0.12	50.0
	3.0	0.14	0.09	55.6
Q345R				
500	1.7	NB	NB	–
	2	NB	NB	–
	2.2	NB	NB	–
	2.6	NB	1.01	–
	3.0	1.18	0.77	53.2
1000	1.7	NB	0.76	–
	2	0.60	0.45	33.3
	2.2	0.48	0.35	37.1
	2.6	0.34	0.25	36.0
	3.0	0.27	0.20	35.0
1500	1.7	0.60	0.46	30.4
	2	0.39	0.27	44.4
	2.2	0.30	0.20	50.0
	2.6	0.20	0.13	53.8
	3.0	0.17	0.10	70.0
ASME SA738 Gr. B				

(continued)

Table 4.2 (continued)

D_i/t	$D_i/2h_i$	Buckling pressure (MPa)		Difference (%)
		Ellipsoidal head	Torispherical head	
500	1.7	NB	NB	–
	2	NB	2.12	–
	2.2	2.21	1.71	29.2
	2.6	1.61	1.30	23.8
	3.0	1.27	1.08	17.6
1000	1.7	1.82	1.44	26.4
	2	1.17	0.87	34.5
	2.2	0.92	0.65	41.5
	2.6	0.62	0.44	40.9
	3.0	0.47	0.33	42.4
1500	1.7	1.20	0.97	23.7
	2	0.77	0.54	42.6
	2.2	0.60	0.40	50.0
	2.6	0.39	0.26	50.0
	3.0	0.28	0.19	47.4

NB = No Buckling

VIII-2 and EN 13445–3 are determined using Formula (1.30). The failure pressures corresponding to yielding of knuckle for EN13445-3 are determined using Formula (1.38) by substituting allowable stress for yield strength. The failure pressures for ASME VIII-1 and GB/T 150.3 are determined using Formulas (1.15) and (1.16) respectively. The failure pressures for JB 4732 are determined using Formula (1.35) for calculating limit pressure.

For example, Heads Nos. ZJU-P5 and ZJU-P15, the geometrical and material parameters of which are presented in Table 3.1 of Sect. 3.2.1, are selected to compare the plastic collapse failure pressures given by the codes and standards and the new Formula (3.2) proposed in Sect. 3.4.2 with experimental plastic collapse pressures. Figure 4.2 shows the failure pressures given by the codes and new Formula (3.2), as well as the experimental results. We can see that among the codes, ASME VIII-2 and EN 13445–3 give maximum failure pressures, but they are 40% lower than the experimental plastic collapse pressures because the effect of material strengthening is not considered, and this leads to conservative design if head thickness is calculated using the current design equations in the codes and standards. However, the failure pressures determined by the new Formula (3.2) show good consistency with the experimental plastic collapse pressures. Therefore, it is recommended that a new design formula which considers strengthening effect of nonlinearity should be proposed for plastic collapse of ellipsoidal heads under internal pressure, as mentioned in Ref. [3].

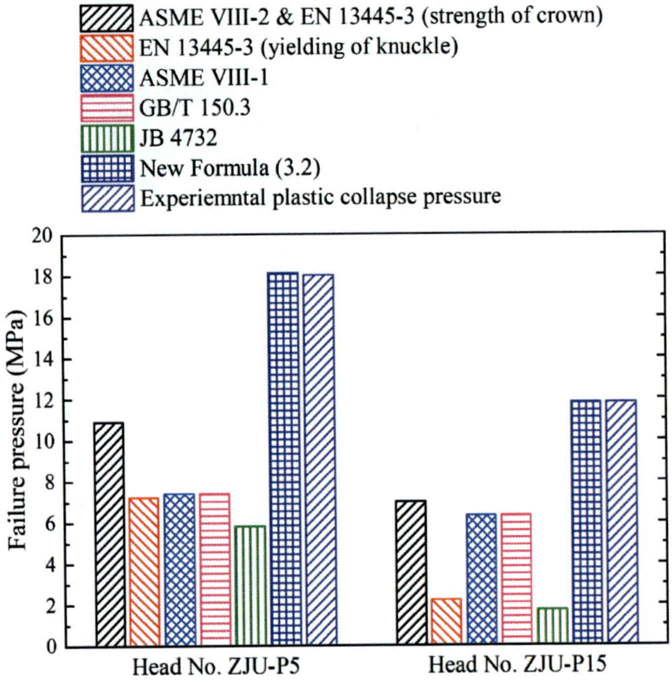

Fig. 4.2 Comparison of failure pressure between codes, new Formula (3.2) and test data

4.2 Failure Mechanism-Based Design Method

As concluded in Sect. 4.1, there are three main problems in the current methods for the design of ellipsoidal heads in the codes and standards: (1) the current buckling criteria are more conservative, limited and/or inaccurate; (2) the current design rules for buckling failure based on geometric equivalency approaches are on the conservative side, resulting in uneconomical design; and (3) the effect of material strengthening is not considered in the current design rules for plastic collapse, also leading to conservative design. In this section, therefore, we will develop a new failure mechanism-based method for the design of ellipsoidal heads under internal pressure.

4.2.1 Applicability Scope

The applicability scope of the new design method is associated with the formulas for predicting the buckling pressure and plastic collapse pressure, as well as the buckling criterion. The new Formula (2.2) for predicting buckling pressure is applicable to ellipsoidal heads with $1.6 \leq D_i/2h_i \leq 3.0$ and $200 \leq D_i/t \leq 2000$, and made of steel with a yield strength of 100–600 MPa. The new buckling criterion (Fig. 2.51) is applicable to ellipsoidal heads with $1.0 \leq D_i/2h_i \leq 3.0$ and $200 \leq D_i/t \leq 2000$, and made of steel with a yield strength of 100–600 MPa. The new Formula (3.2) for predicting plastic collapse is applicable to ellipsoidal heads with $1.5 \leq D_i/2h_i \leq 2.5$ and $20 \leq D_i/t \leq 2000$, and for ellipsoidal heads made of steel including stainless steel, carbon steel and low-alloy steel specified in the pressure vessel codes and standards. Furthermore, ellipsoidal heads having smaller or larger diameter-to-thickness ratio ($D_i/t < 20$ or $D_i/t > 2000$) and smaller or larger radius-to-height ratio ($D_i/2h_i < 1.5$ or $D_i/2h_i > 2.5$) are barely used in practice. Therefore, the applicability scope of the new method for the design of ellipsoidal heads under internal pressure is given as follows:

$$20 \leq \frac{D_i}{t} \leq 2000$$

$$1.5 \leq \frac{D_i}{2h_i} \leq 2.5$$

4.2.2 Formula for Preventing Local Buckling

In Sect. 2.5.2.2, we developed a new formula for predicting buckling pressure of ellipsoidal heads, as given by $P_b = 56\beta S_y\left(\frac{t}{D_i}\right)^{1.29}\left(\frac{2h_i}{D_i}\right)^{1.93}$; see Formula (2.2). $\beta = 1.0$ for ellipsoidal heads fabricated by pressing, spinning, and machining, while $\beta = 0.625$ for ellipsoidal heads assembled from formed segments, a difference which accounts for the large initial shape imperfections arising in this type of head. The predictions of the new formula for buckling pressure are in reasonably good agreement with the experimental results of the tested heads. Compared with other formulas, the new formula has comprehensive advantages in both accuracy and applicability.

Based upon the above prediction formula, a design formula for preventing the buckling of ellipsoidal heads can be developed which is modified by a new factor β_c; see Formula (4.1). For ellipsoidal heads fabricated by pressing, spinning and machining, $\beta_c = 30$ is recommended, assuring that the ratios of experimental buckling pressure to allowable pressure are greater than or equal to 1.5 for these ellipsoidal heads (Table 4.3). Safety factor 1.5 is used by ASME VIII-2 and EN 13345–3

Table 4.3 Comparison of experimental buckling pressures (P_{exp}) with allowable pressures (P) given by Formula (4.1)

Head no	Nominal dimension			Material			P_{exp} (MPa)	P (MPa)	P_{exp}/P
	D_i(mm)	D_i/t	$D_i/2h_i$	Designation	S_y (MPa)	S_u (MPa)			
Cold pressing									
ZJU-P22	1800	600	2.0	S30408	385	689	1.30	0.79	1.6
ZJU-P20	1800	600	2.0	S30408	385	689	1.68	0.79	2.1
ZJU-P18	1800	600	2.0	S31603	310	562	1.43	0.64	2.2
ZJU-P24	1200	480	2.0	SUS3042B	444	775	1.81	1.22	1.5
ZJU-P25	1200	480	2.0	SUS3042B	444	775	2.0	1.22	1.6
ZJU-P21	1200	400	2.0	S30408	385	689	3.63	1.33	2.7
ZJU-P12	1200	480	2.0	316L2B	335	623	2.07	0.92	2.3
Cold spinning									
ZJU-S3	1800	450	2.2	S30408	392	708	2.51	0.97	2.6
ZJU-S5	1800	450	2.4	S30408	392	708	1.77	0.82	2.2
ZJU-S6	1200	400	2.4	S31603	340	605	1.47	0.83	1.8
ZJU-S4	1200	400	2.4	S30408	327	700	1.45	0.8	1.8
RA1	500	500	2.5	Carbon steel	310	340	0.80	0.52	1.5
RA2	500	250	2.5	Carbon steel	300	330	2.50	1.24	2
RA3	500	500	2.5	Stainless steel	480	690	1.50	0.81	1.9
RA4	500	250	2.5	Stainless steel	520	700	4.90	2.15	2.3

(continued)

Table 4.3 (continued)

Head no	Nominal dimension			Material			P_{exp} (MPa)	P (MPa)	P_{exp}/P
	D_i (mm)	D_i/t	$D_i/2h_i$	Designation	S_y (MPa)	S_u (MPa)			
Machined from solid billet									
Gal1	137	1080	2	Mild steel	242	242	0.83	0.23	3.6
Gal2	137	1080	2	Mild steel	242	242	0.76	0.23	3.3
Assembled from formed segments									
ZJU-FA1	4797	872	1.728	SA738 Gr. B	509	602	1.15	0.66	1.7
ZJU-FA2	4797	872	1.728	SA738 Gr. B	588	662	1.15	0.76	1.5
ZJU-FT1	5000	909	2	SA738 Gr. B	612	674	0.70	0.56	1.3
ZJU-FT2	5000	909	2	SA738 Gr. B	558	626	0.70	0.51	1.4
CBI1	4877	1024	2	SA516 Gr. 70	353	552	0.40	0.28	1.4
CBI2	4877	768	2	SA516 Gr. 70	361	538	0.73	0.41	1.8

for formulating the protection against local buckling of internally pressurized ellipsoidal heads (Sect. 1.4.3). For ellipsoidal heads assembled from formed segments, it is recommended that β_c is determined by dividing 56β by the safety factor of 1.5, i.e., $\beta_c = 56 \times 0.625 \div 1.5 = 23$, considering that large shape imperfection in this type of head leads to a considerable decrease in buckling pressure. Therefore, with respect to buckling failure, we developed the new design Formula (4.1) for determining the allowable pressure of ellipsoidal heads under internal pressure.

$$P = \beta_c S_y \left(\frac{t}{D_i}\right)^{1.29} \left(\frac{2h_i}{D_i}\right)^{1.93} \tag{4.1}$$

Regarding the test data on buckling of tested ellipsoidal heads (Table 2.7), safety margin analysis was performed for the new design Formula (4.1). The allowable pressures (P) of the tested heads are determined by Formula (4.1) where the yield strength (S_y) is assumed to be the measured yield strength. The ratios of experimental buckling pressure to allowable pressure (P_{exp}/P) are also given to analyze the safety margins for all tested heads. Table 4.3 shows the experimental buckling pressure, allowable pressure and the ratio P_{exp}/P for the tested heads. We can see that $P_{exp}/P \geq 1.5$ for all tested heads fabricated by pressing, spinning and machining. For ellipsoidal heads assembled from formed segments, $P_{exp}/P \geq 1.5$ for the ZJU-FA1, ZJU-FA2 and CBI2 heads. Therefore, the new design Formula (4.1) for preventing buckling shows sufficient safety margins for the tested heads. For the ZJU-FT1, ZJU-FT2 and CBI1 heads, the shape imperfections in the three heads do not meet the requirements specified in the codes and standards (Sect. 1.5.2.1), causing a significant reduction in buckling pressure, and the ratios of P_{exp}/P are lower than 1.5.

4.2.3 Formula for Preventing Plastic Collapse

In Sect. 3.4.2, we used the finite element method to calculate the plastic collapse pressure of ellipsoidal heads, taking into account the effects of strain hardening and geometric strengthening, and we found that ultimate tensile strength and thickness-to-diameter ratio significantly affect plastic collapse pressure. Thus, the simple Formula (3.2), $P_c = 3.7\frac{t}{D_i}S_u$, for predicting the plastic collapse pressure of ellipsoidal heads was proposed by curve fitting for many finite element results. The plastic collapse pressures of the simple prediction formula show satisfactory agreement with the experimental results of 33 full-scale ellipsoidal heads, covering diameter-to-thickness ratios of 91–1024, radius-to-height ratios of 1.7–2.4, many kinds of steel types, and three kinds of common manufacturing methods (cold pressing, cold spinning, and assembly by formed segments).

Based on the prediction Formula (3.2), a new design equation for preventing the plastic collapse of ellipsoidal heads can be developed which is modified by a new factor α, and plastic collapse pressure (P_c) and ultimate tensile strength (S_u)

are changed into allowable pressure (P) and allowable stress (S) respectively; see Formula (4.2). In order to consider uncertain factors affecting safety, the factor α should be smaller than 3.7 and is determined assuring that the ratios of experimental plastic collapse pressure to allowable pressure are more than or equal to safety factor of 2.4 as used for plastic collapse in ASME VIII-2. For ellipsoidal heads with no weld joints, $\alpha = 3.13$ is recommended. For heads with weld joints, it is recommended that α should be reduced to 2.63 because weld joints may result in a decrease in mechanical properties, leading to a reduction in the plastic collapse pressure of the ellipsoidal heads. Therefore, we developed the new design Formula (4.2) for determining allowable pressure in order to prevent plastic collapse of ellipsoidal heads under internal pressure.

$$P = \alpha \frac{t}{D_{\mathrm{i}}} S \tag{4.2}$$

Regarding the test data on plastic collapse of internally pressurized ellipsoidal heads (Table 3.1), the allowable stress (S) was assumed to be determined by dividing the measured ultimate tensile strength by the safety factor of 2.4; hence, the allowable pressures (P) of the tested heads are determined by Formula (4.2), and the ratios of experimental plastic collapse pressure to allowable pressure (P_{cb}/P) are also given to analyze safety margin for all tested heads. Tables 4.4 and 4.5 present the experimental plastic collapse pressures, allowable pressures and P_{cb}/P ratios for tested heads without and with weld joints respectively. We can see that $P_{\mathrm{cb}}/P \geq 2.4$ for all the tested heads, except the three heads (ZJU-P23, ZJU-P24 and ZJU-P25) made of SUS3042B. In case of these three heads, the rupture occurred in the weld joint between the head and cylinder or in the weld joints in head crown. It can be supposed that the weld joints of SUS3042B were poor, causing a reduction in the load carrying capacity of the heads, and the P_{cb}/P ratios are relatively low but greater than or equal to 2. Therefore, the design Formula (4.2) for preventing plastic collapse shows sufficient safety margins for all tested heads.

4.2.4 *New Design Method for Ellipsoidal Heads Under Internal Pressure*

Based on Formulas (4.1) and (4.2) for determining the allowable pressure, we can obtain the formulas for determining the minimum required thickness, which is usual in engineering application. Thus, the minimum required thickness to prevent the buckling of ellipsoidal heads is given by

$$t_{\mathrm{b}} = D_{\mathrm{i}} \left[\frac{p_{\mathrm{c}}}{\beta_{\mathrm{c}} S_{\mathrm{y}}} \left(\frac{D_{\mathrm{i}}}{2h_{\mathrm{i}}} \right)^{1.93} \right]^{0.77} \tag{4.3}$$

Table 4.4 Comparison of experimental plastic collapse pressures (P_{cb}) with allowable pressures (P) given by Formula (4.2) for heads without weld joints

Head No	Dimension			Material			Rupture position	P_{cb} (MPa)	P (MPa)	P_{cb}/P
	D_i(mm)	D_i/t	$D_i/2h_i$	Designation	S_y (MPa)	S_u (MPa)				
Pressed heads made of carbon and low-alloy steel										
ZJU-P1	500	91	2	SA738B	615	673	Crown	25.5	9.6	2.7
ZJU-P2	600	100	2	Q245R	332	466	Crown	16.9	6.1	2.8
ZJU-P3	600	100	2	16MnDR	399	573	Crown	20.2	7.5	2.7
ZJU-P4	600	100	2	Q345R	414	565	Crown	20	7.4	2.7
ZJU-P5	600	100	2	09MnNiDR	376	500	Crown	18	6.5	2.8
ZJU-P6	1200	200	2	Q245R	332	466	Crown	8.1	3	2.7
ZJU-P7	1200	200	2	Q345R	414	565	Crown	10.3	3.7	2.8
ZJU-P8	1200	200	2	16MnDR	399	573	Crown	10.1	3.7	2.7
ZJU-P9	1800	300	2	Q345R	414	565	Crown	6.8	2.5	2.7
ZJU-P10	1800	300	2	16MnDR	431	575	Crown	6.7	2.5	2.7
ZJU-P11	1800	300	2	Q245R	332	466	Crown	5.7	2	2.9
Pressed heads made of stainless steel										
ZJU-P13	600	200	2	S22053	653	803	Crown	13	5.2	2.5
ZJU-P14	600	200	2	S30403	282	710	Crown	12.7	4.6	2.8
ZJU-P15	600	200	2	S30408	269	680	Crown	11.8	4.4	2.7

(continued)

Table 4.4 (continued)

Head No	Dimension			Material			Rupture position	P_{cb} (MPa)	P (MPa)	P_{cb}/P
	D_i(mm)	D_i/t	$D_i/2h_i$	Designation	S_y (MPa)	S_u (MPa)				
ZJU-P16	600	200	2	S32168	230	610	Crown	10.9	4	2.7
ZJU-P19	600	200	2	S30408	385	689	Head-cylinder weld joint	11	4.5	2.4
ZJU-P23	600	300	2	SUS3042B	402	739	Head-cylinder weld joint	6.7	3.2	2.1
Spun heads with $D_i/2h_i \neq 2$										
ZJU-S1	1200	200	2.4	Q345R	405	547	Crown	8.8	3.6	2.4
ZJU-S2	1200	200	1.7	Q345R	405	547	Crown	8.8	3.6	2.4

Table 4.5 Comparison of experimental plastic collapse pressures (P_{cb}) with allowable pressures (P) given by Formula (4.2) for heads with weld joints

Head No	Dimension			Material			Rupture position	P_{cb}(MPa)	P(MPa)	P_{cb}/P
	D_i(mm)	D_i/t	$D_i/2h_i$	Designation	S_y(MPa)	S_u(MPa)				
Pressed heads made of stainless steel										
ZJU-P12	1200	480	2	316L2B	335	623	Crown	4.1	1.4	2.9
ZJU-P17	1200	300	2	S30408	382	653	Head-cylinder weld joint	8	2.4	3.3
ZJU-P18	1800	600	2	S31603	310	562	Head-cylinder weld joint	3	1	3
ZJU-P20	1800	600	2	S30408	385	689	Head-cylinder weld joint	3.2	1.3	2.5
ZJU-P21	1200	400	2	S30408	385	689	Head-cylinder weld joint	5	1.9	2.6
ZJU-P22	1800	600	2	S30408	385	689	Head-cylinder weld joint	3.1	1.3	2.4
ZJU-P24	1200	480	2	SUS3042B	444	775	Weld joint in crown	4.1	1.8	2.3
ZJU-P25	1200	480	2	SUS3042B	444	775	Weld joint in crown	3.6	1.8	2
Spun heads with $D_i/2h_i \neq 2$										
ZJU-S3	1800	450	2.2	S30408	392	708	Weld joint	5.1	1.7	3
ZJU-S4	1200	400	2.4	S30408	327	700	Weld joint	5.9	1.9	3.1
ZJU-S5	1800	450	2.4	S30408	392	708	Weld joint	5.1	1.7	3
ZJU-S6	1200	400	2.4	S31603	340	605	Weld joint	6	1.7	3.5
Heads assembled from formed segments										
CBI1	4877	1024	2	SA516 Gr. 70	353	552	Weld joint in knuckle	1.6	0.6	2.7
CBI2	4877	768	2	SA516 Gr. 70	361	538	Weld joint in crown	2.3	0.8	2.9

where β_c is 30 for ellipsoidal heads fabricated by pressing, spinning and machining, and 23 for ellipsoidal heads assembled by formed segments. And the minimum required thickness to prevent the plastic collapse of ellipsoidal heads is given by

$$t_c = \alpha_c \frac{p_c D_i}{S} \tag{4.4}$$

where $\alpha_c = 1/\alpha$; thus, α_c is 0.32 for ellipsoidal heads without weld joints, and 0.38 for ellipsoidal heads with weld joints.

In addition, the new buckling criterion presented in Sect. 2.4.4 is used to determine whether buckling failure is considered or not in the head design. We changed the curve of the buckling criterion (Fig. 2.51) into the corresponding formula in the range $1.5 \le D_i/2h_i \le 2.5$; see Formula (4.5) [1]. If the values of $D_i/2h_i$ and D_i/t_c meet the requirements of Formula (4.5), buckling failure shall be considered in head design, otherwise it shall not be considered.

$$D_i/2h_i \ge \begin{cases} 1.514 + 71.593\exp\left(-\frac{D_i/t_c+65}{57.951}\right) + 1.647\exp\left(-\frac{D_i/t_c+65}{340.700}\right), 200 \le D_i/t_c < 935 \\ 1.6, 935 \le D_i/t_c \le 2000 \end{cases} \tag{4.5}$$

Therefore, we developed a new failure mechanism-based design method for ellipsoidal heads under internal pressure. Figure 4.3 presents the design chart of the new method: First, the calculation pressure (p_c), head inside diameter (D_i), radius-to-height ratio ($D_i/2h_i$) and allowable stress (S) and yield strength (S_y) of the material are determined according to design conditions. Second, the minimum required thickness (t_c) to prevent plastic collapse is determined using Formula (4.4). Third, Formula (4.5) is used to determine whether buckling is considered or not. If buckling is not considered, the required thickness of the ellipsoidal heads is finally determined as the minimum required thickness (t_c) to prevent plastic collapse. If buckling is considered, the minimum required thickness (t_b) to prevent buckling is determined by using Formula (4.3). Finally, the required thickness of ellipsoidal heads is determined as the maximum value between the required thickness (t_c) to prevent plastic collapse and required thickness (t_b) to prevent buckling.

As stated in Sect. 1.4.1, the design method based on limit analysis in the pre-2004 revision of ASME VIII-2 is still used for the design of ellipsoidal heads in Chinese industry standard JB 4732; however, the post-2007 revision of ASME VIII-2 abandoned the design method. Recently, JB 4732 was revised, and this design method based on limit analysis was also deleted. The conservative modification for the above new method herein was made and approved by China Standardization Committee on Boilers and Pressure Vessels: regarding Formula (4.3) for preventing buckling, β_c is only 23 for all ellipsoidal heads, whether pressing, spinning or assembly by formed segments; regarding Formula (4.4) for preventing plastic collapse, α_c reduces to 0.42 for ellipsoidal heads without nozzles, and $\alpha_c = 0.45$ for ellipsoidal heads with

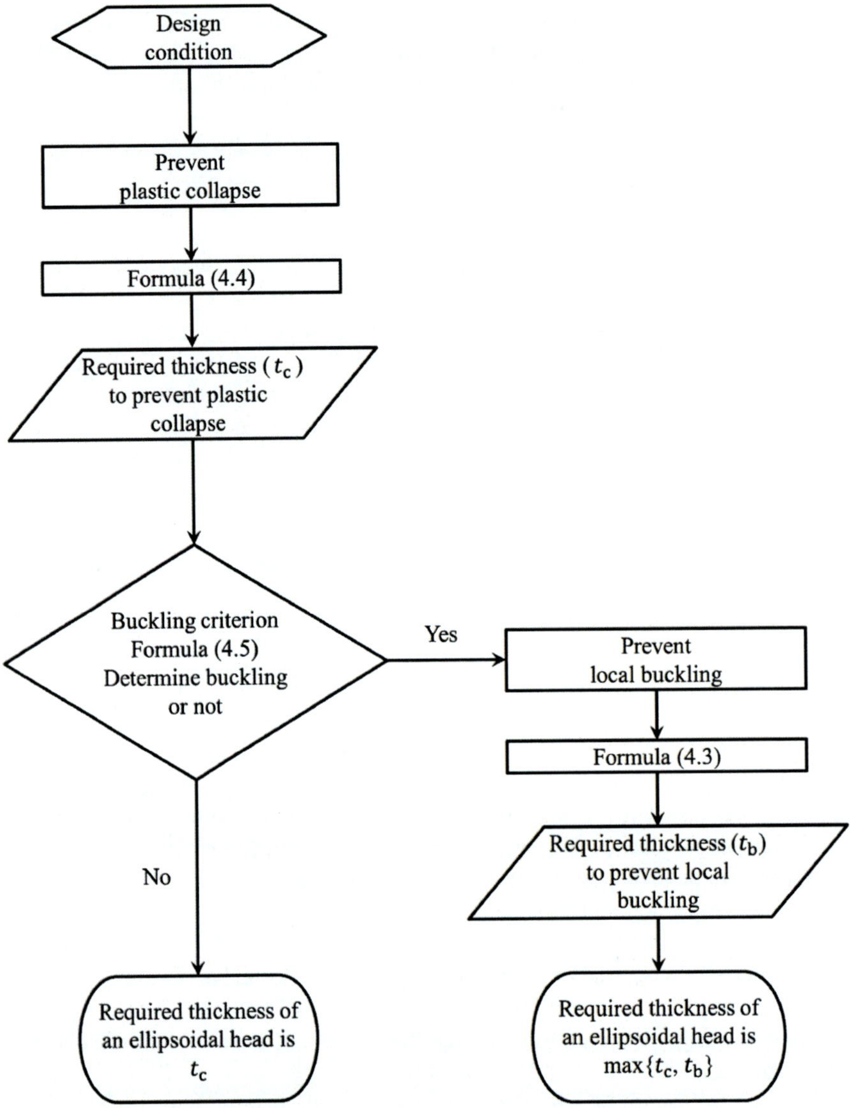

Fig. 4.3 Flow chart for the design of ellipsoidal heads under internal pressure

nozzles [4]; and the buckling criterion of Formula (4.5) is reduced to $D_i/t_c > 200$. Research on the modified revision of the new method is seen in the thesis of Zhang [5].

Table 4.6 Summary of applicability scope

Method name	Applicability scope
EN 13445–3	$12.5 \leq D_o/t \leq 1000$, $1.785 \leq D_i/2h_i < 2.2$
ASME VIII-2	$20 \leq L/t \leq 2000$; $1.7 \leq D_i/2h_i \leq 2.2$ For usual ellipsoidal heads with $D_i/2h_i = 2$, $L = 0.9D_i$, thus $22 \leq D_i/t \leq 2222$
New design method	$20 \leq D_i/t \leq 2000$; $1.5 \leq D_i/2h_i \leq 2.5$

4.3 Comparison Between New Method and Those in Standards

4.3.1 Comparison of Applicability Scope

The applicability scope of the method for the design of ellipsoidal heads is provided in EN 13445–3 and ASME VIII-2. The applicability scope of EN 13445–3 and ASME VIII-2, and the new design method herein are summarized in Table 4.6. In EN13445, the scope of the radius-to-height ratio is $1.7 < D_i/2h_i < 2.2$; however, the scope of $D_i/2h_i$ is revised as $1.785 < D_i/2h_i < 2.2$ due to the limitation of $r/D_i \leq 0.2$ according to Formula (1.25). In ASME VIII-2, the scope of the crown radius-to-thickness ratio is $20 \leq L/t \leq 2000$ for torispherical heads; according to Formula (1.26), the crown radius-to-thickness ratio scope of torispherical heads can be changed into the diameter-to-thickness ratio scope of ellipsoidal heads. In the case of usual ellipsoidal heads with $D_i/2h_i = 2$, $L = 0.9D_i$ according to Formula (1.26), thus the diameter-to-thickness ratio scope is $22 \leq D_i/t \leq 2222$ for ellipsoidal heads with $D_i/2h_i = 2$. We can see that the applicability scope of the new design method is wider than those of ASME VIII-2 and EN 13445–3, specially the scope of radius-to-height ratio.

4.3.2 Comparison of Required Thickness

It is assumed that ellipsoidal heads have an inside diameter of 2000 mm, and three radius-to-height ratios ($D_i/2h_i$) are included in the applicability scope of the codes (standards) and the new design method (Sect. 4.2): 1.8, 2.0 and 2.19. Ellipsoidal heads have no weld joints and are fabricated by pressing or spinning. In addition, the investigated ellipsoidal heads are assumed to be made of carbon steel with a Young's modulus of 200 GPa, yield strength of 300 MPa and allowable stress of 180 MPa. The allowable stress is determined by dividing the ultimate tensile strength by design factor according to the specifications in codes and standards. The required thicknesses of ellipsoidal heads under different calculation pressures are calculated by the design methods in the current codes and standards ASME VIII-1, ASME

VIII-2, GB/T 150.3, JB 4732 and EN 13445–3 presented in Sect. 1.4, as well as the new design method herein.

Table 4.7 and Fig. 4.4 show a comparison of required thickness between the current codes (standards) and new design method for seamless ellipsoidal heads made of carbon steel. Two types of failure, collapse and buckling, are considered in ASME VIII-1 and VIII-2. At lower calculation pressure (p_c), the required thickness to prevent buckling is greater than the thickness to prevent collapse because buckling is a governing failure for thin heads under lower pressure conditions. As the calculation pressure increases, collapse becomes a governing failure, so the required thickness of the heads is determined as the required thickness to prevent collapse. The required thickness to prevent buckling is very close between ASME VIII-1 and VIII-2 because the design method for preventing buckling in the two ASME codes is determined from Miller's formulas. The required thickness to prevent collapse in ASME VIII-2 is smaller than ASME VIII-1 because the design methods for preventing collapse in the two ASME codes are different: the method in ASME VIII-2 is based on the rupture of the equivalent head crown corresponding to torispherical heads, and the method in ASME VIII-1 is based on the maximum stress criterion associated with the stress ratio (K), as mentioned in Sect. 1.4.2. For EN 13445, three types of failure are considered: crown collapse, knuckle yielding and buckling. At lower calculation pressure, buckling controls the required thickness of the heads. As the calculation pressure increases, the required thickness to prevent knuckle yielding becomes greater than the thickness to prevent crown collapse, which is the same as ASME VIII-2, but the thickness to prevent crown collapse is greater than the required thickness to prevent knuckle yielding for heads with $D_i/2h_i$ of 1.8 when the calculation pressure is much higher. Buckling is not considered in GB/T 150.3 and JB 4732, and the applicability scope of JB 4732 is limited to $t/L \geq 0.002$ when the calculation pressure is much lower; thus, some super-thin heads are not designed using the methods in GB/T150.3 and JB 4732. Similar to ASME VIII-1 and VIII-2, the new design method herein considers two types of failure: collapse and buckling. Different from the current codes and standards, the required thickness to prevent collapse does not change with the radius-to-height ratio because the plastic collapse pressure of ellipsoidal heads has no relationship with the radius-to-height ratio in the range of $1.7 \leq D_i/2h_i \leq 2.2$; see Sect. 3.3.3. With respect to the current design methods in the codes and standards, the required thickness given by JB 4732 is the greatest, while ASME VIII-2 provides the smallest results for most of heads. The required thickness given by the new method is more than 20% smaller than ASME VIII-2, showing that the new method provides more economical design results compared with the current design methods.

Table 4.7 Required thickness comparison between current codes and new design method for seamless ellipsoidal heads made of carbon steel

p_c (MPa)	$D_i/2h_i$	Required thickness t (mm)														
		ASME VIII-1			ASME VIII-2			EN 13445–3				GB/T 150.3	JB 4732	New method		
		t_c (Collapse)	t_b (Buckling)	max $\{t_c, t_b\}$	t_c (Collapse)	t_b (Buckling)	max $\{t_c, t_b\}$	t_c (Collapse)	t_y (Yielding)	t_b (Buckling)	max $\{t_c, t_y, t_b\}$	t_c (Collapse)	t_c (Collapse)	t_c (Collapse)	t_b (Buckling)	max $\{t_c, t_b\}$
0.5	1.8	2.4	2.4	2.4	2.3	2.4	2.4	2.3	2.4	3.1	3.1	2.4	–	1.8	2.5	2.5
0.75	1.8	3.6	2.9	3.6	3.4	3	3.4	3.4	3.6	4.1	4.1	3.6	4.5	2.7	3.5	3.5
1.5	1.8	7.3	–	7.3	6.8	–	6.8	6.8	7.2	6.5	7.2	7.3	8.6	5.3	–	5.3
3	1.8	14.6	–	14.6	13.6	–	13.6	13.6	14	–	14	14.6	16.3	10.7	–	10.7
5	1.8	24.3	–	24.3	22.7	–	22.7	22.7	22.7	–	22.7	24.4	25.9	17.8	–	17.8
7	1.8	34.1	–	34.1	31.9	–	31.9	31.9	31.7	–	31.9	34.3	35.1	24.9	–	24.9
10	1.8	48.8	–	48.8	45.7	–	45.7	45.7	45.2	–	45.7	49.2	48.2	35.6	–	35.6
0.5	2	2.8	3	3	2.5	3	3	2.5	3.4	3.7	3.7	2.8	4	1.8	3	3
0.75	2	4.2	–	4.2	3.8	3.7	3.8	3.8	5	4.8	5	4.2	5.8	2.7	4	4
1.5	2	8.3	–	8.3	7.5	–	7.5	7.5	9.4	7.7	9.4	8.4	10.9	5.3	6.9	6.9
3	2	16.7	–	16.7	15.1	–	15.1	15.1	17.4	–	17.4	16.7	20.1	10.7	–	10.7
5	2	27.9	–	27.9	25.2	–	25.2	25.2	27.7	–	27.7	28	31.3	17.8	–	17.8
7	2	39	–	44.6	35.3	–	35.3	35.3	37.9	–	37.9	39.3	41.7	24.9	–	24.9
10	2	55.9	–	55.9	50.7	–	50.7	50.7	52.3	–	52.3	56.3	56.2	35.6	–	35.6
0.5	2.19	3.1	3.4	3.4	2.7	3.4	3.4	2.7	4.3	4.3	4.3	3.1	4.9	1.8	3.4	3.4

(continued)

Table 4.7 (continued)

p_c (MPa)	$D_i/2h_i$	Required thickness t (mm)														
		ASME VIII-1			ASME VIII-2			EN 13445-3				GB/T 150.3	JB 4732	New method		
		t_c (Collapse)	t_b (Buckling)	max $\{t_c, t_b\}$	t_c (Collapse)	t_b (Buckling)	max $\{t_c, t_b\}$	t_c (Collapse)	t_y (Yielding)	t_b (Buckling)	max $\{t_c, t_y, t_b\}$	t_c (Collapse)	t_c (Collapse)	t_c (Collapse)	t_b (Buckling)	max $\{t_c, t_b\}$
0.75	2.19	4.7	–	4.7	4.1	4.1	4.1	4.1	6.2	5.6	6.2	4.7	7	2.7	4.6	4.6
1.5	2.19	9.4	–	9.4	8.2	6	8.2	8.2	11.4	-	11.4	9.5	13.1	5.3	7.9	7.9
3	2.19	18.9	–	18.9	16.5	–	16.5	16.5	20.5	-	20.5	19	23.7	10.7	–	10.7
5	2.19	31.6	–	31.6	27.5	–	27.5	27.5	32.1	-	32.1	31.7	36.4	17.8	–	17.8
7	2.19	44.2	–	44.2	38.6	–	38.6	38.6	43.3	-	43.3	44.5	48	24.9	–	24.9
10	2.19	63.3	–	63.3	55.4	–	55.4	55.4	58.6	-	58.6	63.8	63.9	35.6	–	35.6

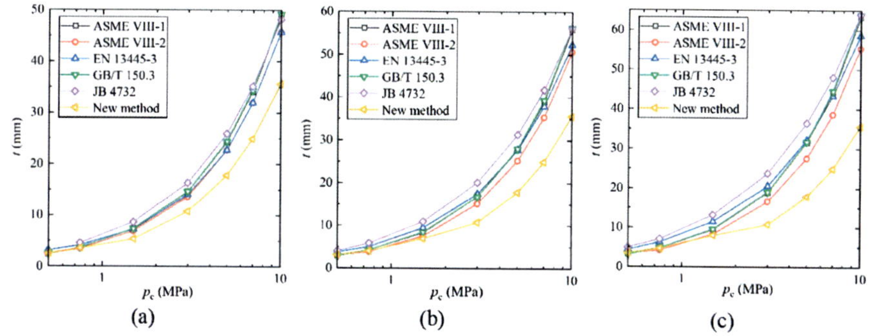

Fig. 4.4 Required thickness comparison between current codes and new design method for seamless ellipsoidal heads made of carbon steel: **a** $D_i/2h_i = 1.8$; **b** $D_i/2h_i = 2$; **c** $D_i/2h_i = 2.19$

It is assumed that seamless ellipsoidal heads with a diameter of 1000 mm are made of stainless steel with a Young's modulus of 195 GPa, yield strength of 180 MPa and allowable stress of 120 MPa. The allowable stress is determined by dividing the yield strength by design factor of 1.5. Table 4.8 and Fig. 4.5 show a comparison of required thickness between the current codes and new design method for seamless ellipsoidal heads made of stainless steel. It can be seen that the required thickness given by the new method is smallest for most of the heads as compared with the current codes and standards.

In addition, Table 4.9 and Fig. 4.6 present required thickness comparisons between the current codes (standards) and new design method for welded ellipsoidal heads with a diameter of 2000 mm which are made of the above stainless steel and fabricated by pressing and/or spinning. Table 4.10 and Fig. 4.7 present required thickness comparisons between the current codes (standards) and new design method for welded ellipsoidal heads with a diameter of 6000 mm which are made of the above carbon steel and fabricated by segments. We can also see that for most of the welded ellipsoidal heads, the new method still provides smaller required thicknesses than those of the current codes and standards. Therefore, the new method for ellipsoidal heads provides more economical design results as compared with the current codes and standards.

4.3.3 Advantages of New Method

Compared with the current design methods in codes and standards, the new method for the design of ellipsoidal heads has the following advantages: (1) The applicability scope of the new design method is wider than those of ASME VIII-2 and EN 13445–3; (2) The new local buckling criterion is more accurate and reasonable than the others, making it more applicable to determine whether internally pressurized ellipsoidal heads will buckle or not; and (3) The new method can provide smaller

Table 4.8 Required thickness comparison between current codes and new design method for seamless ellipsoidal heads made of stainless steel

p_c(MPa)	$D_i/2h_i$	Required thickness t (mm)						
		ASME VIII-1	ASME VIII-2	EN 13445–3 (Pressing)	EN 13445–3 (Spinning)	GB/T 150.3	JB 4732	New method
0.5	1.8	1.8	1.7	2.2	1.8	1.8	2.3	1.9
0.75	1.8	2.7	2.5	2.9	2.7	2.7	3.3	2
1.5	1.8	5.5	5.1	5.3	5.3	5.5	6.3	4
3	1.8	10.9	10.2	10.2	10.2	11	11.8	8
5	1.8	18.3	17.1	17.1	17.1	18.4	18.7	13.3
7	1.8	25.6	24	24	24	25.8	25.2	18.7
10	1.8	36.7	34.6	34.6	34.6	37.2	34.5	26.7
0.5	2	1.6	1.9	2.6	1.9	2.1	2.9	2.2
0.75	2	3.1	2.8	3.6	3.6	3.1	4.2	3
1.5	2	6.3	5.6	6.8	6.8	6.3	7.8	4
3	2	12.5	11.3	12.6	12.6	12.6	14.3	8
5	2	20.9	18.9	20.2	20.2	21.1	22.1	13.3
7	2	29.3	26.6	27.3	27.3	29.6	29.3	18.7
10	2	42	38.3	38.3	38.3	42.6	39.3	26.7
0.5	2.19	2.4	2.1	3.1	3.1	2.4	3.5	2.5
0.75	2.19	3.5	3.1	4.4	4.4	3.5	5.1	3.4
1.5	2.19	7.1	6.2	8	8	7.1	9.3	4
3	2.19	14.2	12.4	14.6	14.6	14.2	16.7	8
5	2.19	23.7	20.7	23	23	23.8	25.4	13.3
7	2.19	33.2	29.1	30.5	30.5	33.5	33.2	18.7
10	2.19	47.6	41.9	41.9	41.9	48.2	44	26.7

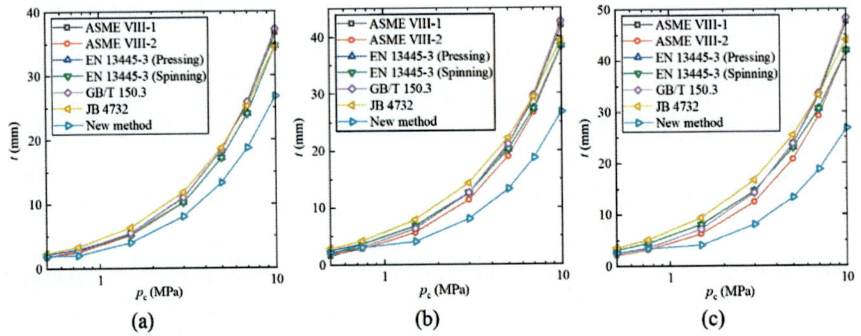

Fig. 4.5 Required thickness comparison between current codes and new design method for seamless ellipsoidal heads made of stainless steel: **a** $D_i/2h_i = 1.8$; **b** $D_i/2h_i = 2$; **c** $D_i/2h_i = 2.19$

Table 4.9 Required thickness comparison between current codes and new design method for welded ellipsoidal heads fabricated by pressing or spinning

p_c(MPa)	$D_i/2h_i$	Required thickness t (mm)					
		ASME VIII-1	ASME VIII-2	EN 13445–3	GB/T 150.3	JB 4732	New method
0.25	1.8	1.8	1.9	2.8	1.8	–	2.2
0.5	1.8	3.6	3.4	4.4	3.6	4.5	3.8
0.75	1.8	5.5	5.1	5.8	5.5	6.6	4.8
1.5	1.8	10.9	10.2	10.6	11	12.6	9.5
3	1.8	21.9	20.4	20.5	22	23.6	19
5	1.8	36.5	34.2	34.2	36.8	37.3	31.7
7	1.8	51.2	48.1	48.1	51.7	50.4	44.3
10	1.8	73.4	69.1	69.1	74.3	69	63.3
0.25	2	2.3	2.3	3.3	2.1	–	2.6
0.5	2	4.2	3.8	5.2	4.2	5.8	4.4
0.75	2	6.3	5.6	7.3	6.3	8.4	6
1.5	2	12.5	11.3	13.5	12.5	15.6	9.5
3	2	25.1	22.6	25.1	25.2	28.6	19
5	2	41.8	37.9	40.3	42.1	44.2	31.7
7	2	58.7	53.3	54.6	59.2	58.6	44.3
10	2	84	76.6	76.6	85.1	78.6	63.3
0.25	2.19	2.6	2.6	3.8	2.4	–	2.9
0.5	2.19	4.7	4.1	6.2	4.7	7	5
0.75	2.19	7.1	6.2	8.8	7.1	10.1	6.9
1.5	2.19	14.2	12.3	16	14.2	18.6	9.5
3	2.19	28.4	24.7	29.2	28.5	33.4	19
5	2.19	47.4	41.4	46	47.7	50.7	31.7
7	2.19	66.5	58.2	61.1	67.1	66.4	44.3
10	2.19	95.2	83.7	83.7	96.4	88	63.3

required thicknesses than the current codes and standards including ASME VIII-1, ASME VIII-2, GB/T 150.3, JB 4732 and EN 13445–3, which gives it an economical advantage because the effects of material hardening and geometric strengthening are considered.

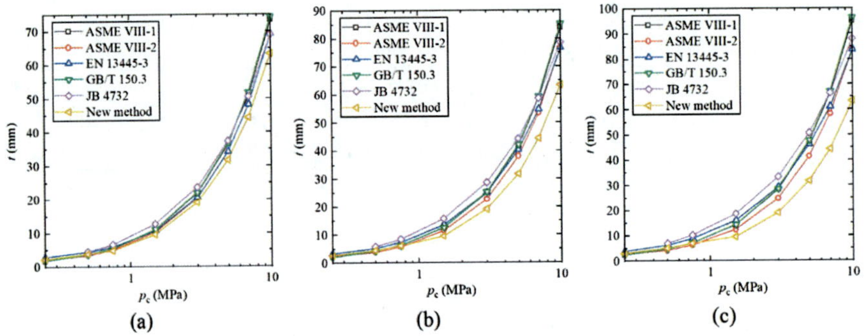

Fig. 4.6 Required thickness comparison between current codes and new design method for welded ellipsoidal heads fabricated by pressing or spinning: **a** $D_i/2h_i = 1.8$; **b** $D_i/2h_i = 2$; **c** $D_i/2h_i = 2.19$

Table 4.10 Required thickness comparison between current codes and new design method for welded ellipsoidal heads fabricated by segments

p_c(MPa)	$D_i/2h_i$	Required thickness t (mm)					
		ASME VIII-1	ASME VIII-2	EN 13445–3	GB/T 150.3	JB 4732	New method
0.11	1.8	2.9	3	3.4	1.6	–	2.9
0.25	1.8	4.8	4.8	5.9	3.6	–	5.5
0.5	1.8	7.3	7.1	9.4	7.3	9.3	9.3
0.75	1.8	10.9	10.2	12.4	10.9	13.6	12.7
1.5	1.8	21.9	20.3	21.5	21.9	25.9	19
3	1.8	43.7	40.8	41.9	43.8	49	38
5	1.8	73	68.1	68.2	73.3	77.8	63.3
7	1.8	102.3	95.7	95.7	102.9	105.3	88.7
10	1.8	146.4	137.2	137.2	147.6	144.7	126.7
0.11	2	3.9	3.9	4	1.8	–	3.4
0.25	2	6.2	6.2	7	4.2	–	6.4
0.5	2	9	9	11.1	8.3	11.9	10.9
0.75	2	12.5	11.3	15	12.5	17.3	14.9
1.5	2	25	22.5	28.3	25.1	32.6	19
3	2	50.1	45.2	52.3	50.2	60.3	38
5	2	83.6	75.5	83	83.9	93.9	63.3
7	2	117.1	106	113.6	117.8	125	88.7
10	2	167.6	152.1	156.9	169	168.7	126.7
0.11	2.19	4.7	4.6	4.7	2.1	–	3.9
0.25	2.19	7.2	7.24	8.1	4.7	–	7.3
0.5	2.19	10.1	10.2	12.8	9.4	14.6	12.5

(continued)

Table 4.10 (continued)

p_c(MPa)	$D_i/2h_i$	Required thickness t (mm)					
		ASME VIII-1	ASME VIII-2	EN 13445–3	GB/T 150.3	JB 4732	New method
0.75	2.19	14.2	12.3	18.5	14.2	21.1	17.1
1.5	2.19	28.3	24.6	34	28.4	39.2	29.1
3	2.19	56.7	49.4	61.5	56.9	71.2	38
5	2.19	94.7	82.5	96.4	95.1	109.2	63.3
7	2.19	132.7	115.9	129.9	133.4	143.9	88.7
10	2.19	189.8	166.2	175.9	191.4	191.7	126.7

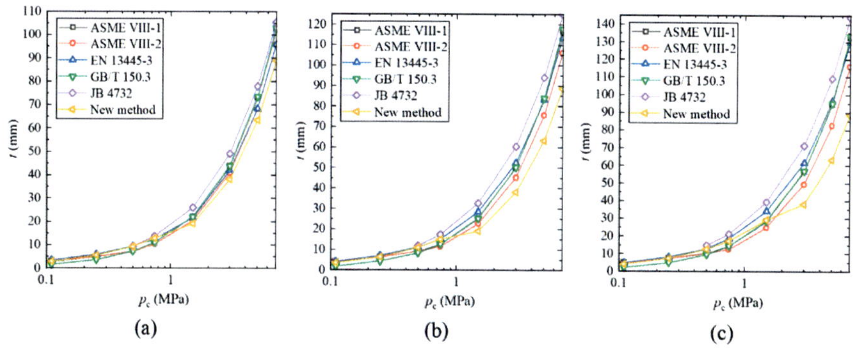

Fig. 4.7 Required thickness comparison between current codes and new design method for welded ellipsoidal heads fabricated by segments: **a** $D_i/2h_i = 1.8$; **b** $D_i/2h_i = 2$; **c** $D_i/2h_i = 2.19$

References

1. Li K (2019) Research on buckling behavior and prediction method of large-scale thin-walled ellipsoidal head under internal pressure. PhD Thesis, Zhejiang University, Hangzhou. (in Chinese)
2. Galletly GD (1981) A comparison of the plastic buckling behaviour of 2:1 ellipsoidal and 2:1 torispherical shells subjected to internal pressure. In: Hult J, Lemaitre J (eds) Physical non-linearities in structural analysis, international union of theoretical and applied mechanics. Springer, Berlin
3. Zheng J, Yu Y, Chen Y, Li K, Zhang Z, Peng W, Gu C, Xu P (2021) Comparison of ellipsoidal and equivalent torispherical heads under internal pressure: buckling, plastic collapse and design rules. ASME J Press Vessel Technol 143:021301
4. Li K, Zheng J, Zhang Z, Gu C, Xu P (2020) A new method for preventing plastic collapse of ellipsoidal head under internal pressure. In: ASME PVP conference, Paper No. PVP2020–21212.
5. Zhang Z (2021) Research on failure mechanism-based method for design of steel ellipsoidal heads under internal pressure. PhD Thesis, Zhejiang University, Hangzhou.

Chapter 5
Control of Fabrication Quality of Ellipsoidal Heads

Generally, fabrication quality affects the load carrying capacity or in-service performance of pressure components; for example, shape deviation arises during head fabrication, and head buckling is sensitive to shape deviation [1–3]. Forming strain also occurs during head fabrication, where plastic deformation tends to cause phase transformation in materials, which significantly affects material properties [4]. In particular, metastable austenitic stainless steel (m-ASS) can undergo strain-induced martensitic transformation (SIM-Tr) during cold forming of heads, which may cause degradation in mechanical properties of m-ASS at cryogenic temperature [4–8]. Forming temperature is a key operation parameter for warm forming process of m-ASS heads as SIM-Tr is also sensitive to forming temperature except forming strain [9–11]. Therefore, it is essential to control head fabrication quality for designing ellipsoidal heads. This section mainly focuses on controlling fabrication quality of ellipsoidal heads, including shape deviation, forming strain, and forming temperature.

5.1 Effects of Fabrication on Head Performance

5.1.1 Effects of Shape Deviation on Buckling Pressure

As mentioned in Sect. 2.4.1, we compared the buckling pressures of the tested ellipsoidal head models with perfect shape and actual shape obtained by 3D scanner measurement, which showed that the buckling pressures of the head models with actual shape were lower than those with perfect shape (Table 2.4). As presented in Sect. 5.2.2, a bulging of weld, a common shape deviation, arises along the meridian of the knuckle in the fabrication of large ellipsoidal heads. We generated a FE head model with a bulging of weld (see Sect. 2.4.3.5) to investigate the effects of bulging height on the buckling pressure of an internally pressurized ellipsoidal

J. Zheng and K. Li, *New Theory and Design of Ellipsoidal Heads for Pressure Vessels*,
https://doi.org/10.1007/978-981-16-0467-6_5

head. Figure 5.1 shows the buckling pressures of head models with different bulging heights, and it can be seen that buckling pressure decreases with the increase in bulging height. This section is from our papers [12, 13].

In addition, we investigated the effects of shape deviation on buckling pressure of ellipsoidal heads under external pressure. In the case of the tested head ZJU-FT1, we performed eigenvalue buckling analysis on the ZJU-FT1 head models with perfect and actual shapes. Figure 5.2 shows the buckling modes of the ZJU-FT1 head models

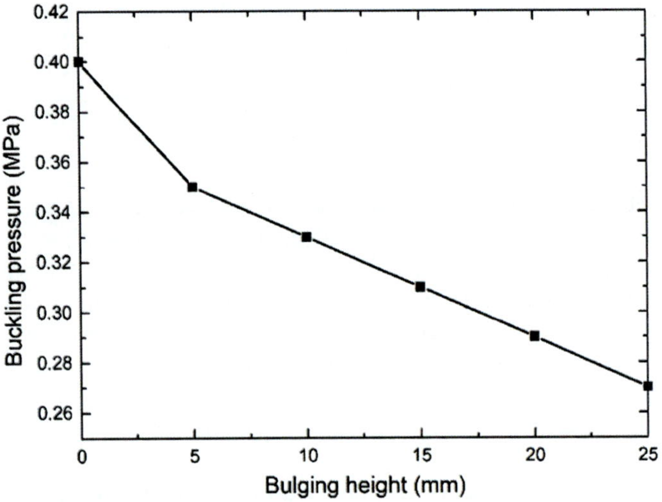

Fig. 5.1 Effects of bulging height on buckling pressure of ellipsoidal heads under internal pressure (from Ref. [13])

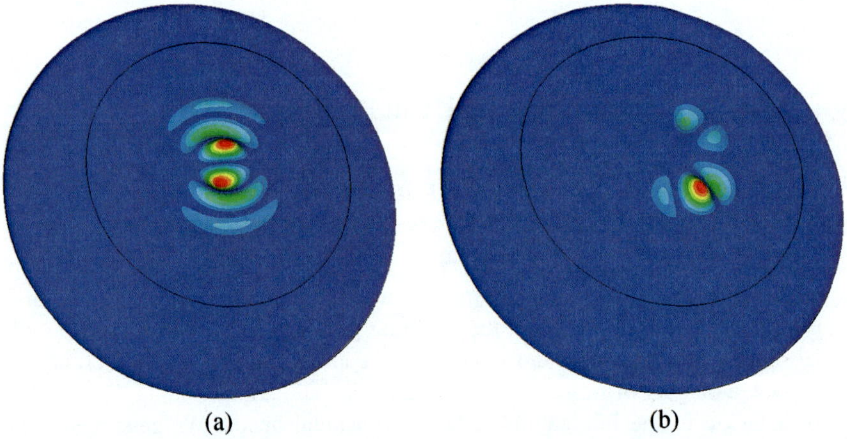

Fig. 5.2 Buckling modes of the ZJU-FT1 head models under external pressure for **a** perfect shape and **b** actual shape

under external pressure, and the eigenvalue buckling pressures are determined as 0.35 MPa and 0.16 MPa for perfect shape and actual shape respectively. Obviously, the buckling pressure of the ZJU-FT1 head model with actual shape is much lower than that with perfect shape. Therefore, shape deviation has a significant effect on the buckling pressure of ellipsoidal heads under internal/external pressure: shape deviation leads to a reduction in buckling pressure.

5.1.2 Effects of Forming Strain on Mechanical Properties

To investigate the effects of forming strain on the mechanical properties of head material such as austenitic stainless steel S30408, we performed 35% uniaxial pre-strain at 20 °C on an as-received test piece of S30408 plate with a thickness of 14 mm, which is equivalent to forming strain in heads during stamping, while ignoring the difference in stress state. A tensile specimen labeled CPS35% was cut from the 35% pre-strained test piece, while a tensile specimen labeled ASR was cut from the as-received test piece for comparison. Considering that ellipsoidal heads made of S30408 are usually used for the closure of cryogenic vessels, cryogenic tensile properties were tested at −196 °C on the two tensile specimens CPS35% and ASR. The tensile test was performed according to the Chinese standards GB/T 13239 [14] and GB/T 228.1. Figure 5.3 shows the cryogenic tensile test results, including yield strength (0.2% proof stress $R_{p0.2}$, MPa), ultimate tensile strength (S_u, MPa), elongation to fracture (A, %), and reduction of area (Z, %).

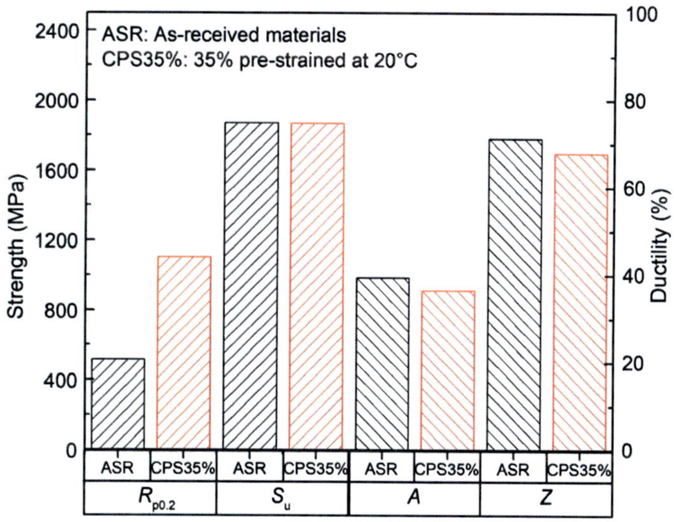

Fig. 5.3 Comparison of cryogenic tensile test results between as-received and 35% pre-strained S30408 plates

According to Fig. 5.3, the cryogenic yield strength ($R_{p0.2}$) of the pre-strained specimen CPS35% soars by approximately 114% compared to the ASR results, while the ultimate tensile strength (S_u) of CPS35% basically remains the same compared to the ASR results, with a difference lower than 1%. For cryogenic ductility, the elongation to fracture (A) of CPS35% is lower than that of ASR by nearly 8%, while the reduction of area (Z) of CPS35% is lower than that of ASR by less than 5%. The significant enhancement in yield strength after the performance of forming strain can be explained by the work hardening effect due to dislocation strengthening and martensite transformation. Thus, the influence of forming strain on cryogenic tensile properties shows a significant strengthening effect in yield strength.

We subsequently investigated the effects of forming strain on material impact properties. Similar to the tensile specimens, Charpy V-notch specimens were cut from the as-received test piece of the S30408 plate and 35% pre-strained test piece. The machining of the Charpy V-notch pendulum impact test pieces was performed according to the specifications of Chinese standard GB/T 229 [15]. In addition, to study the effects of forming strain caused by stamping processing on the mechanical properties of ellipsoidal heads, we also carried out a cold stamping test on a type EHA350 × 6 head, which was a 2:1 ellipsoidal head with a diameter of 350 mm and thickness of 6 mm. The type EHA350 × 6 head was stamped at room temperature using a 630-ton double-action hydraulic oil press machine. The material for the head stamping test was m-ASS S30408 treated with solid solution. Charpy V-notch specimens were cut from the type EHA350 × 6 head labeled CSH along the meridional direction, with the distance from the head edge to the joint of the head knuckle and crown covering around 70 mm. The 70 mm was divided into six even lateral sampling spaces, and three specimens were machined at each lateral sampling space, as shown in Fig. 5.4.

The cryogenic impact test was carried out using a ZBC2302-D pendulum impact test machine with a maximum pendulum energy of 450 J. The specimens were pre-cooled in liquid nitrogen for 30 min before the tests. The values of absorbed energy (KV_2) and lateral expansion (LE), which represent the material impact properties, were measured, as shown in Fig. 5.5a and b respectively.

As shown in Fig. 5.5, both the impact energy and lateral expansion values of the CPS material from the 35% pre-strained plate are much lower than those of the as-received material (ASR), showing that forming strain causes a degradation in cryogenic impact properties. Compared to the as-received material (ASR), both the impact energy and lateral expansion values of the material from the EHA350 × 6 head (CSH) are lower due to the forming strain occurring in the knuckle and straight sections of the head after stamping. For the CPS material, the degradation of cryogenic impact properties is basically between the mean values of the CSH material in the head knuckle section and the straight section. For the CSH material, the cryogenic impact properties degrade more severely with the increase in distance from head apex along the meridional direction; that is, the impact properties of the knuckle section are higher than those of the straight section. For absorbed energy (KV_2), although the mean KV_2 value of the knuckle section is higher than that of the straight section by approximately 64%, it is still nearly 76% of the mean value

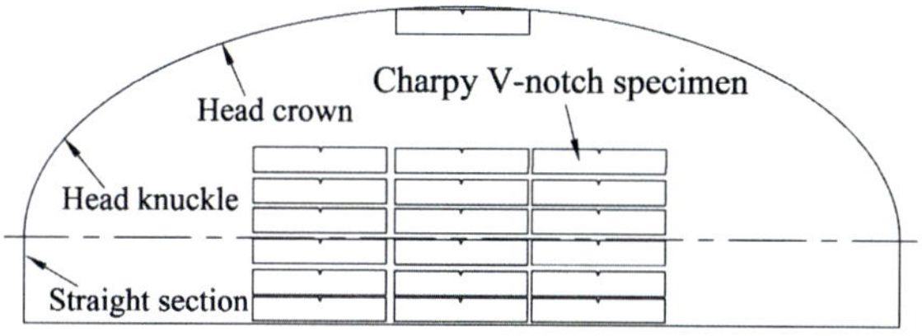

(a)

(b)

Fig. 5.4 **a** Illustration of Charpy V-notch impact test piece sampling; **b** EHA350 × 6 head after sampling (from Ref. [16])

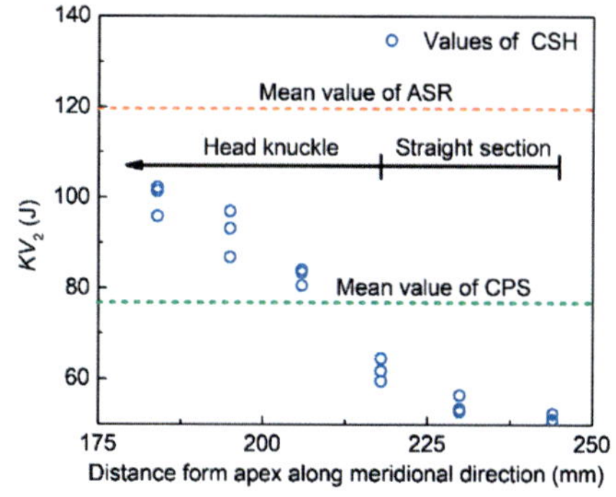

(a)

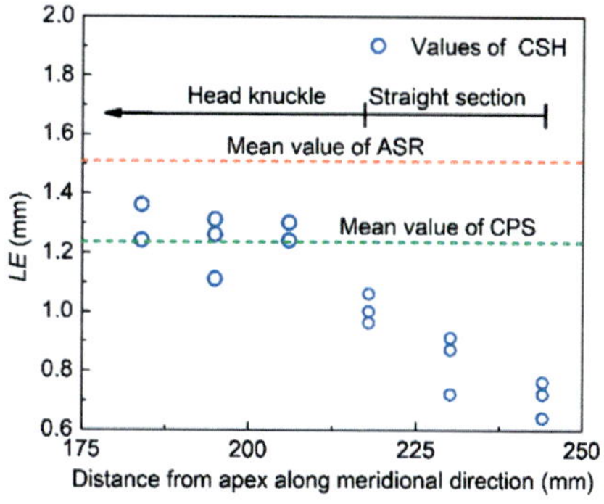

(b)

Fig. 5.5 Cryogenic impact tests results of as-received plate (ASR), 35% pre-strained plate (CPS), and materials machined from EHA350 × 6 head (CSH): **a** Absorbed energy (KV_2); **b** Lateral expansion (LE) (from Ref. [21] with permission of ASME) of the as-received material. It is also similar for the lateral expansion (*LE*) values, the mean *LE* value of the knuckle is higher than that of the straight section by approximately 48%, it is still nearly 84% of the mean *LE* value of the ASR. This can be explained by the plastic deformation in the straight section being larger than the knuckle section, not to mention the ASR. Thus, attention should be paid to the degradation of cryogenic impact properties under the influence of forming strain.

5.1.3 Effects of Forming Temperature on Mechanical Properties

Similarly, to investigate the influence of forming temperature on the mechanical properties of austenitic stainless steel S30408, 35% uniaxial pre-strain was performed at 95 °C on as-received test pieces, which was considered equivalent to forming strain in heads during warm stamping. Similar to the 35% pre-strained S30408 plate at 20 °C (CPS35%) presented in Sect. 5.1.2, cryogenic tensile properties were tested at −196 °C on the 35% pre-strained S30408 plate at 95 °C (WPS35%). Figure 5.6 shows a comparison of the cryogenic tensile test results between 35% pre-strained S30408 plates at 20 °C (CPS35%) and 95 °C (WPS35%).

With the comparison of cryogenic tensile properties results of CPS35% and WPS35% in Fig. 5.6, it is clear that the yield strength ($R_{p0.2}$) of the WPS35% material declines by approximately 23% compared with the CPS35% material, while the ultimate strength (S_u) values remain basically constant. For cryogenic ductility, the

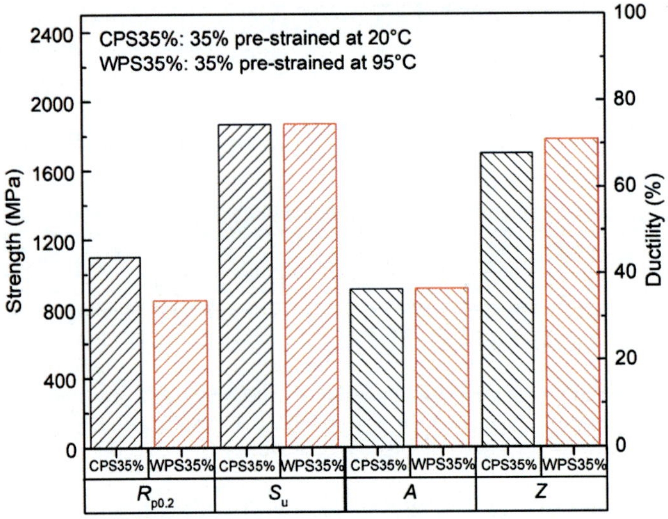

Fig. 5.6 Comparison of cryogenic tensile test results between 35% pre-strained S30408 plates at 20 °C (CPS35%) and 95 °C (WPS35%)

A and Z values of the WPS35% material are both slightly improved by less than 5% compared with the CPS35% material. Although the yield strength degrades with the increase in forming temperature from 20 to 95 °C, it is still enhanced by nearly 64% compared to the ASR material (Fig. 5.3).

To study the effects of forming temperature on the mechanical properties of ellipsoidal heads, we also carried out a warm stamping test at temperatures ranging from 90 to 120 °C on an EHA350 × 6 head (WSH) which had the same geometrical parameters and material as a cold stamped head EHA350 × 6 (CSH) for comparison. The temperature of the plate during stamping was measured with an infrared thermometer, and a welding torch was used as required to warm up the straight section of WSH. Similar to CSH, impact tests were conducted for WSH at a cryogenic temperature of −196 °C.

Figure 5.7 shows the cryogenic impact test results (absorbed energy KV_2 and lateral expansion LE) of the knuckle and straight sections of WSH and CSH. Regarding the material in the knuckle section, the absorbed energy and lateral expansion of WSH change very little compared with those of CSH. For the material in the straight section, both the absorbed energy and lateral expansion of WSH are higher than those of CSH: the average KV_2 value of WSH is approximately 19% higher than that of CSH, and the average LE value of WSH are significantly higher than that of CSH by approximately 38%. This indicates that an m-ASS ellipsoidal head formed by warm stamping can greatly enhance the impact properties of head material, especially in the straight section.

Sections 5.1.2 and 5.1.3 are from Ref. [16].

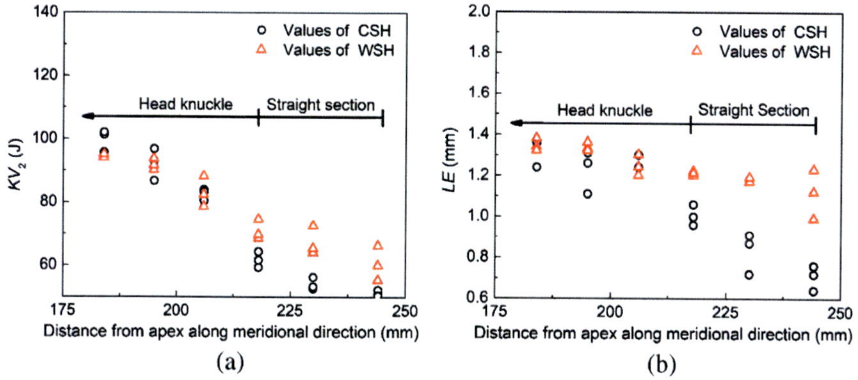

Fig. 5.7 Cryogenic impact tests results of knuckle and straight sections of WSH and CSH heads: **a** Absorbed energy (KV_2); **b** Lateral expansion (LE) (from Ref. [21] with permission of ASME)

5.2 Shape Deviation

As concluded in Sect. 5.1.1, shape deviation leads to a reduction in the buckling pressure of ellipsoidal heads under internal/external pressure. Therefore, we must control head shape quality to prevent the buckling failure of actual ellipsoidal heads. In this section, we performed non-contact measurement of the shape deviation of ellipsoidal heads, studied the characterization of shape deviation, and proposed a method for evaluating shape deviation based on non-contact measurement.

5.2.1 Non-contact Measurement of Shape Deviation

Generally, a wooden template is specially manufactured for measuring the shape deviation of ellipsoidal heads, as shown in Fig. 5.8. The template is rotated about the head axis, and shape deviation is measured along the different head meridians. However, it is not easy to measure the shape deviation of large heads using such a template, and measuring accuracy is hard to guarantee.

Using a 3D laser scanner has become a priority for shape measurement in recent years. As mentioned in Sect. 2.2.2, we measured the shape of the tested ellipsoidal heads with a diameter of 5000 mm using a terrestrial 3D scanner, which is highly applicable for measuring the shape of large-scale structures (Fig. 2.12). For small heads with a diameter of less than 5000 mm, it is recommended to use a portable 3D sensor to measure head shape, as shown in Fig. 5.9. The portable 3D scanner is a highly accurate non-contact measuring instrument with a measurement resolution of 0.025 mm, maximum accuracy of 0.025 mm, and stand-off distance of 300 mm. After non-contact measurement using 3D laser scanners, we obtained the measured shapes of the ellipsoidal heads. The shape deviation could then be determined by comparing the measured shape with the perfect shape.

Fig. 5.8 Measurement of shape deviation of ellipsoidal head using template

Fig. 5.9 Shape measurement of ellipsoidal head using portable 3D scanner (from Ref. [35])

5.2.2 *Characterization of Shape Deviation*

Figure 5.10 shows the shape deviations of the tested ellipsoidal heads fabricated by pressing, spinning and assembly by segments. For the ZJU-FA1 head assembled from segments, the maximum outside shape deviation Δ_o was 50.6 mm (i.e. Δ_o/D_n = 1.05%) and the maximum inside shape deviation (Δ_i) was 34.5 mm (i.e. Δ_i/D_n = 0.719%), as shown in Fig. 5.10a. For the pressed ZJU-P24 head with a nominal diameter (D_n) of 1200 mm, the maximum outside shape deviation (Δ_o) is 3.2 mm (i.e. $\Delta_o/D_n = 0.27\%$) and the maximum inside shape deviation (Δ_i) was 3.8 mm (i.e. $\Delta_i/D_n = 0.32\%$), as shown in Fig. 5.10b. For the spun ZJU-S6 head with a nominal diameter (D_n) of 1200 mm, as shown in Fig. 5.10c, the maximum outside shape deviation Δ_o was 6.4 mm (i.e. $\Delta_o/D_n = 0.53\%$) and the maximum inside shape deviation (Δ_i) was 7.2 mm (i.e. $\Delta_i/D_n = 0.6\%$), as also shown in Fig. 5.11 for shape deviation along the meridian of the spun ZJU-S6 head. It can be seen that the order of shape deviation of ellipsoidal heads is: heads assembled from segments > spun heads > pressed heads, depending on the fabrication method (see Sect. 1.5.1).

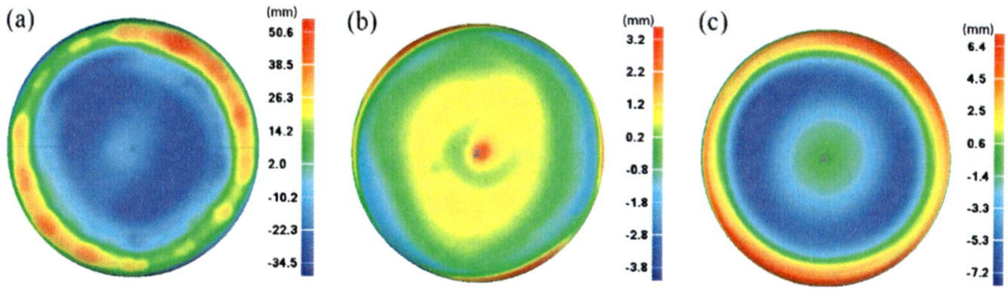

Fig. 5.10 Characterization of shape deviation based on 3D scanner measurement: **a** Head assembled from segments (ZJU-FA1); **b** Pressed head (ZJU-P24); **c** Spun head (ZJU-S6) (from Ref. [13])

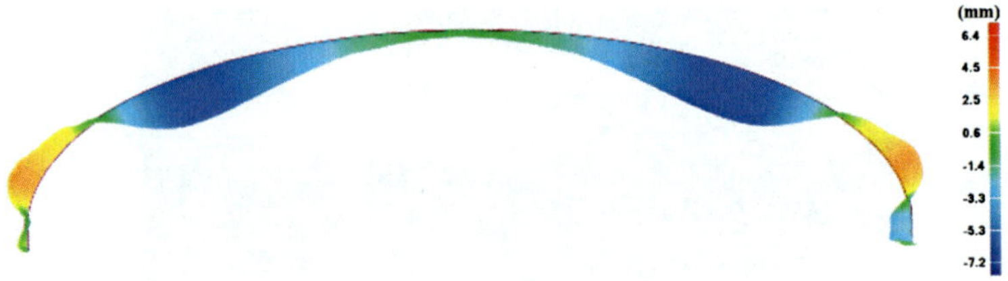

Fig. 5.11 Shape deviation along meridian of spun head (ZJU-S6)

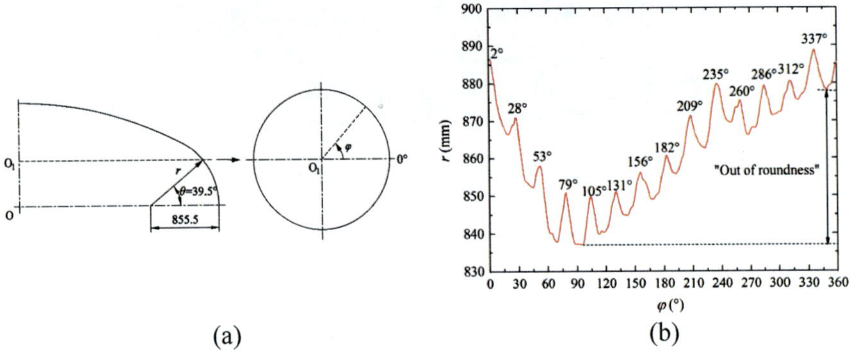

Fig. 5.12 Measured radius distribution of head knuckle along circumferential line: **a** Location of circumferential line; **b** Radius r versus circumferential angle φ (from Ref. [12] with permission of Elsevier)

Pressed heads are formed by a die with an ellipsoid shape, so the shape deviation of pressed ellipsoidal heads is the minimum. For spun heads and heads assembled from segments, the crowns and knuckles are formed separately, so they are approximate ellipsoidal heads or torispherical heads. Ellipsoidal heads assembled from segments have more weld joints causing bulges, so they have the maximum shape deviation.

Figure 5.12 shows the radius distribution of the ZJU-FT1 head knuckle along a circumferential line obtained from the measured overall shape. The overall difference of the radius is caused by "out of roundness", while the bulges arise due to welding (see Fig. 5.13). Bulges are located at the weld joints in the knuckles of the heads assembled from segments (see Fig. 5.12a). Figures 5.14 and 5.15 show the measurements of bulging of weld joint (79°) in the circumferential and meridional directions respectively. The maximum bulging is at the 79° weld joint, with a bulging height of 11.1 mm. The radiuses at the meridional angles 5° and 57.6° in the meridional direction are different due to "out-of-roundness". The bulging between the meridional angles 23.6° and 57.6° in the meridional direction is caused by welding, as in the case of the bulging in the circumferential direction. It can be also seen that radius r is linear with respect to meridional angle θ and circumferential angle φ respectively. Hence, Formula (5.1) was assumed to characterize the bulging of weld in the knuckle of ellipsoidal heads assembled from segments. This section is from Ref. [12] with permission of Elsevier.

Fig. 5.13 Image of bulging of weld in knuckle of heads assembled from segments (from Ref. [12] with permission of Elsevier)

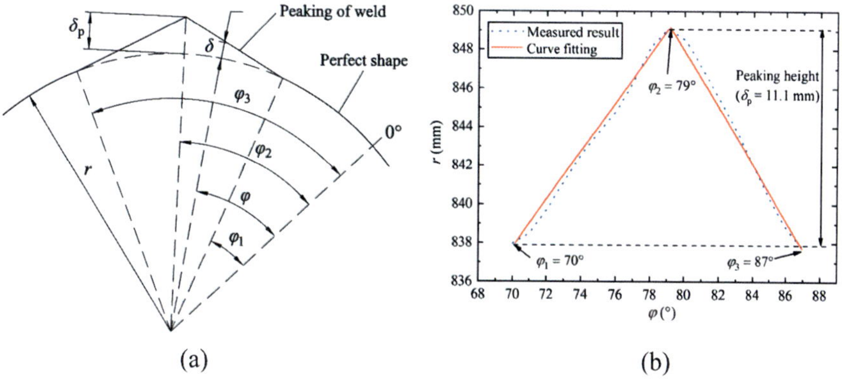

Fig. 5.14 Measurements of bulging of weld joint (79°) in circumferential direction: **a** Bulging of weld; **b** Radius r versus circumferential angle φ (from Ref. [12] with permission of Elsevier)

$$\delta(\varphi,\theta)=\begin{cases}\delta_p\frac{\varphi-\varphi_1}{\varphi_2-\varphi_1}\frac{\theta-\theta_1}{\theta_2-\theta_1}, & \varphi_1\le\varphi<\varphi_2, \theta_1\le\theta<\theta_2\\ \delta_p\frac{\varphi-\varphi_1}{\varphi_2-\varphi_1}\frac{\theta-\theta_2}{\theta_3-\theta_2}, & \varphi_1\le\varphi<\varphi_2, \theta_2\le\theta\le\theta_3\\ \delta_p\frac{\varphi-\varphi_2}{\varphi_3-\varphi_2}\frac{\theta-\theta_1}{\theta_2-\theta_1}, & \varphi_2\le\varphi\le\varphi_3, \theta_1\le\theta<\theta_2\\ \delta_p\frac{\varphi-\varphi_2}{\varphi_3-\varphi_2}\frac{\theta-\theta_2}{\theta_3-\theta_2}, & \varphi_2\le\varphi\le\varphi_3, \theta_2\le\theta\le\theta_3\end{cases} \tag{5.1}$$

where $\delta(\varphi,\theta)$ is the normal deviation from perfect shape, δ_p is bulging height, φ_1, φ_2 and φ_3 are circumferential angles for bulging of weld (see Fig. 5.14), and θ_1, θ_2 and θ_3 are meridional angles for bulging of weld (see Fig. 5.15).

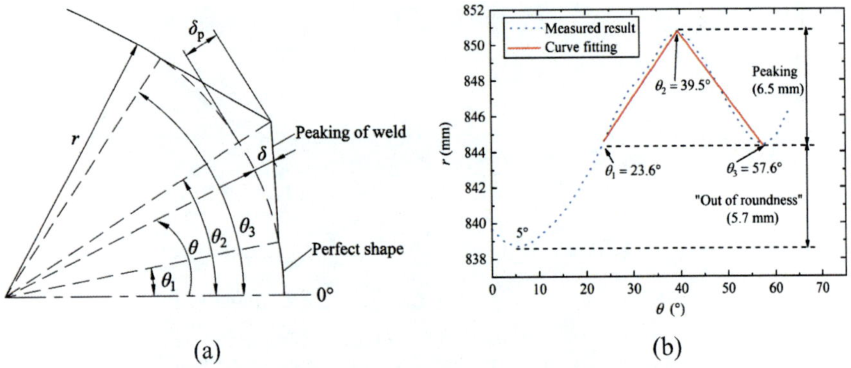

Fig. 5.15 Measurements of bulging of weld joint ($\varphi = 79°$) in meridional direction: **a** Bulging of weld; **b** Radius r versus meridional angle θ (from Ref. [12] with permission of Elsevier)

5.2.3 Evaluation Method of Shape Deviation Using Non-Contact Measurement

The measured shape deviations presented in Sect. 5.2.2 can be evaluated according to pressure vessel codes and standards such as ASME VIII-1 and GB/T 150.4, which specify the shape deviation requirements of ellipsoidal heads: the inner surface of the head shall not deviate from the inside of the specified shape by more than 0.625% of the nominal diameter (i.e. $\Delta_i/D_n \leq 0.625\%$), nor the outside of the specified shape by more than 1.25% of the nominal diameter (i.e. $\Delta_o/D_n \leq 1.25\%$). In the case of the spun ZJU-S6 head with a nominal diameter (D_n) of 1200 mm, we can see from Figs. 5.10c and 5.11 that the maximum outside shape deviation (Δ_o) is 6.4 mm (i.e. $\Delta_o/D_n = 0.53\%$) and the maximum inside shape deviation (Δ_i) is 7.2 mm (i.e. $\Delta_i/D_n = 0.6\%$). Thus, the shape deviations of the ZJU-S6 head meet the requirements specified in ASME VIII-1 and GB/T 150.4.

In summary, we developed a method for evaluating the shape deviation of ellipsoidal heads using non-contact measurement. First, we measured head shape using 3D scanner; a portable 3D scanner is recommended for heads with a diameter of less than 5 m and a terrestrial 3D scanner is recommended for large heads with a diameter of up to tens of meters. Second, the measured shape was compared with the perfect shape, and the overall shape deviation was obtained. Finally, the shape deviation was evaluated according to codes and standards such as ASME VIII-1 and GB/T 150.4.

5.3 Forming Strain

As mentioned in Sect. 5.1.2, forming strain affects the mechanical properties of material, especially by reducing its impact properties; that is, the mechanical properties are associated with forming strain. Heat treatment is subsequently desired at certain conditions to restore the mechanical properties of the head material after forming. According to the pressure vessel codes and standards presented in Sect. 1.5.2.2, forming strain is an essential criterion for deciding whether post-forming heat treatment is required. Therefore, an accurate prediction of the forming strain of heads during the stamping process is a vital aspect of controlling head quality.

In this section, the stamping process of ellipsoidal heads was simulated using the FE method that was experimentally verified by comparing the simulation to the real longitudinal elongation on stamped heads. With the assistance of the FE simulation, the plastic deformation and stress conditions during the stamping process were analyzed. The stamping processes of various sizes of ellipsoidal heads were simulated, and the resultant maximum equivalent plastic strain (PEEQ) was fitted by an empirical formula correlating with the diameters and thicknesses of the heads. The comparison between this empirical formula and other methods suggested by the current codes and standards was discussed. Section 5.3 is from Ref. [17] with permission of Elsevier.

5.3.1 Head Forming Simulation

In order to identify the large plastic deformation distribution on an ellipsoidal head during the stamping process, a reference coordinate was defined on the formed head, as shown in Fig. 5.16a. The X axis represents the longitudinal direction on the outer surface of the head, the θ axis is the latitudinal direction and the δ axis is the thickness direction. The Y axis stands for the symmetric axis of the head, which also shows the stamping direction. The solo X coordinate is sufficient to locate any infinitesimal element of interest due to the well-defined axial-symmetry shape of the formed head. Moreover, similar coordinates denoted with an additional subscript of "0" is also constructed on the blank before stamping, as shown in Fig. 5.16b.

Standard ellipsoidal heads with a major/minor axis ratio of 2 are discussed as they are most widely used in real application. For the convenience of the discussion, the ellipsoidal head was divided into three sections based on the curvature radius of the ellipse. As shown in Fig. 5.17, the blue line represents the ratio of the curvature

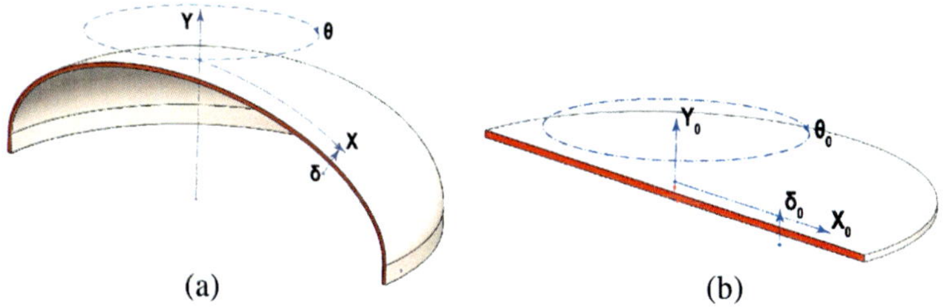

Fig. 5.16 Coordinate definitions for ellipsoidal head stamping: **a** Coordinates on stamped head; **b** Coordinates on blank (from Ref. [17] with permission of Elsevier)

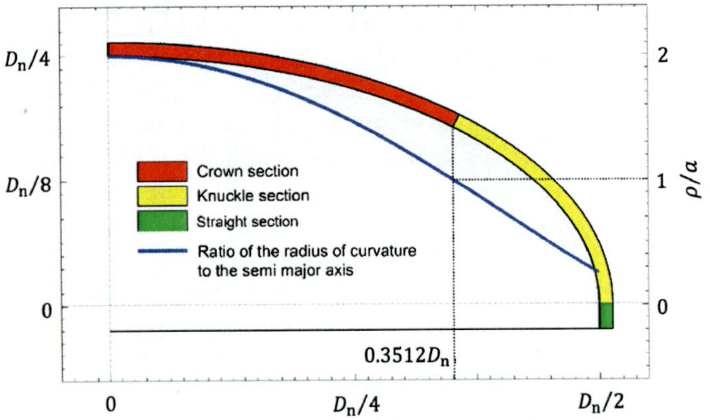

Fig. 5.17 Cross-section view of ellipsoidal head divided into three sections based on curvature radius: (1) Crown section; (2) Knuckle section; (3) Straight section (from Ref. [17] with permission of Elsevier)

radius to the semi major axis (ρ/a), which varies from 2 to 0.25 as its position moves outwards from the center. The "Crown section" is identified as $\rho/a \geq 1$ while the rest part of the ellipse is denoted as the "Knuckle section". The "Straight section" is the straight part at the edge of the head. Because the projection of the inner surface on the cross-section is a pre-defined ellipse with an eccentricity of $\sqrt{3}/2$, by neglecting the offset due to the thickness, such division on the X coordinate (which is set on the outer surface of the head) is approximately given as follows: crown section ($0 \leq X < 0.362D_n$), knuckle section ($0.362D_n \leq X < 0.606D_n$) and straight section ($0.606D_n \leq X \leq 0.606D_n + h_s$), where D_n is the normal diameter of the head, and h_s is the height of the straight section: 25 mm for heads smaller than 2000 mm and 40 mm for larger heads.

As schematically shown in Fig. 5.18, the head forming process by cold stamping consists of the following steps: first (A), the blank is placed on the lubricated die; then (A => B), the binder moves down to press the periphery region of the blank forming a bend outwards over the draw-bead which is applied to prevent detects such as wrinkling, while the punch remains still; next (B => C => D => E), certain force is applied on the binder to clamp the blank and the punch moves down to press the blank and deform it into the desired shape; finally (E => F), the binder and the punch move up, and the formed head is lifted and taken off the machine.

As significant plastic deformation occurs involving material and geometric nonlinearity, FE method was used to simulate the stamping process of the ellipsoidal head. Thanks to the axisymmetric shape of the head and the load/constraint conditions during stamping, only 1/4 of the head and stamping tool was modelled as shown in Fig. 5.19. The model was discretized by the 8-node brick element C3D8R. The die, binder and punch were set to be rigid and meshed with sparse grids because their deformations during the stamping process were negligible comparing with that of the blank, whereas the blank under stamping was meshed with much finer grids as

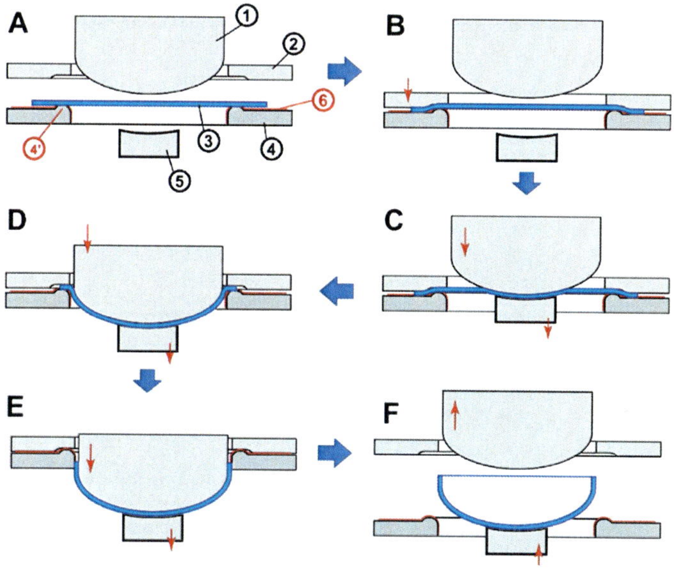

Fig. 5.18 Schematic illustration of head stamping process. 1. Punch; 2. Binder; 3. Blank; 4. Die; 5. Bottom holder; 6. Lubrication layers (from Ref. [17] with permission of Elsevier)

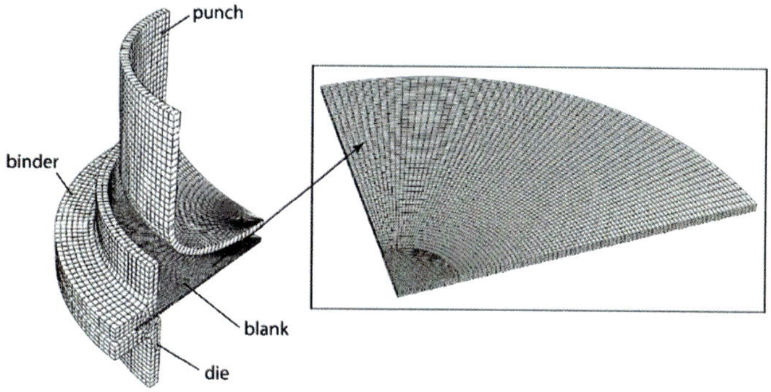

Fig. 5.19 Geometry modeling and meshing in FE simulation (from Ref. [17] with permission of Elsevier)

shown in Fig. 5.19. Furthermore, all degrees of freedom (DOFs) associated with the die were constrained, while the punch and the binder were allowed to move in the Y axis direction only. Furthermore, to simulate the lubrication between the blank and the die/binder, a friction factor of 0.15 was chosen for the stainless steel heads, 0.25 for the carbon steel heads and 0.3 between the blank and the punch.

5.3.2 *Measurement of Forming Strain and Simulation Validation*

In order to validate the FE modeling, extensive measurements of plastic deformation during the stamping process of the ellipsoidal heads were carried out. The detailed experimental procedure and typical measurement results are presented in this section for comparison with the simulations.

5.3.2.1 Materials and Head Size

The head materials under investigation are common commercial grade steels for pressure vessels, including high alloy steel S30408, low alloy steels SA516Gr485 and 16MnDR, and carbon steel Q235B. The chemical compositions of these steels are listed in Table 5.1. The constitutive curves of these materials were obtained via standard tensile tests, and are presented in Fig. 5.20. They constituted the critical input for the FE simulation.

Seven heads of different sizes and materials were measured in this study, and Table 5.2 lists the head specifications and corresponding materials and dimensions.

5.3.2.2 Measurement Method

The measurement took the following steps:

(1) First, perpendicular reference diameter lines were drawn on the blank sheet, as shown in Fig. 5.21a.
(2) Measuring markers were then painted along the reference lines to proper scale. The markers were finer at the edges of the blank sheet (corresponding to the straight section and knuckle section of the formed head) where large strain normally occurs, and sparse at the center of the blank (i.e. the crown section of the head) which hardly deforms.
(3) Next, the blank was stamped into an ellipsoidal head (Fig. 5.21b).
(4) On the formed head, the positions of all the markers (X coordinate) were measured, as shown in Fig. 5.21c.
(5) Finally, the means of the measured values on the four radiuses were calculated, and longitudinal elongation ΔX were calculated as $\Delta X = X_{\text{mean}} - X_0$.

5.3.2.3 Measurement Results and FE Modeling Validation

To verify the FE model, the measured longitudinal elongation ΔX as a function of the X coordinate was compared to the simulated data. By taking two heads, EHA800X6_Q235B and EHA1000X14_S30408, as examples, Fig. 5.22 illustrates the longitudinal elongation distribution along the X coordinate. As the position moves

Table 5.1 Chemical compositions of materials under investigation (%) (from Ref. [17] with permission of Elsevier)

Material	C	Si	Mn	P	S	N	Al	Ni	Cr	Cu	Mo	Ti	Nb	V
S30408[a]	0.037	0.43	1.17	0.026	0.003	0.050	–	8.27	18.12	–	–	–	–	–
Q235B[b]	0.15	0.17	0.31	0.010	0.006	–	–	–	–	–	–	–	–	–
SA516Gr485	0.17	0.26	1.10	0.015	0.008	–	0.026	0.006	0.047	0.057	–	0.006	0.002	0.002
16MnDR[c]	0.16	0.27	1.46	0.013	0.003	–	0.033	0.010	0.020	0.010	0.002	0.015	0.014	0.003

[a] Equivalent to UNS S30400 Grade 304, JIS SUS304, or EN X5CrNi18-10 Nr.1.4301
[b] Equivalent to ASTM A283 Gr.D, EN S235JR, or JIS SS400B
[c] Equivalent to ASME SA516Gr485

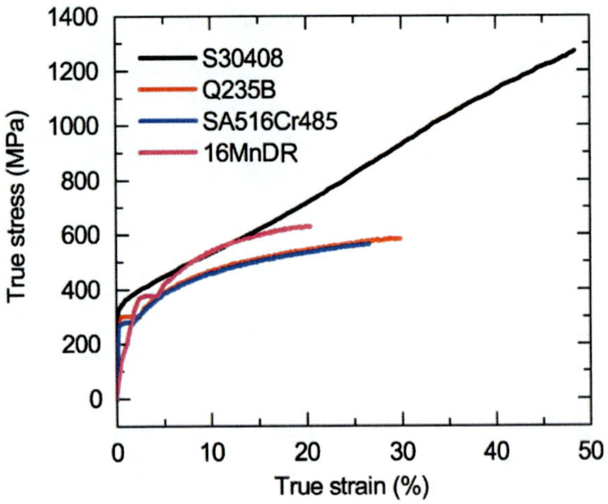

Fig. 5.20 Stress–strain curves of materials under investigation (from Ref. [17] with permission of Elsevier)

Table 5.2 Detailed information of measured heads (from Ref. [17] with permission of Elsevier)

Specification	Material	Diameter, D_n (mm)	Thickness, δ_n (mm)	Height of straight section, h_s (mm)
EHA 800X4	Q235B	800	4	25
EHA 800X6	Q235B	800	6	25
EHA 800X18	SA516Gr485	800	18	25
EHA1000X12	SA516Gr485	1000	12	25
EHA1000X14	S30408	1000	14	25
EHA1200X13	16MnDR	1200	13	25
EHA1800X8	S30408	1800	8	25

from the center to the edge, the longitudinal elongation ΔX increases quickly. The simulated ΔX_{sim} for both heads are lower than the measurement. The root-mean-square deviations (RMSD) between the simulation and measurement were calculated as 2.53 mm and 1.94 mm for EHA1000X14 and EHA800X6 respectively. Nevertheless, the simulation shows a good trend of longitudinal elongation distribution, and the simulated data generally lies in the measurement error bands. The measurement errors here are mainly contributed by two factors: 0.5 mm from the precision of the steel tape and approximately 0.5% from the reading of random errors, such as unpredictable fluctuations in the readings and non-perfect alignment of the steel tape during measurement. Moreover, the constitutive curves of the materials obtained from tensile tests may also introduce errors that result in deviations between the measurement and simulation.

Furthermore, simulations were carried out for all measured heads and the results were compared to the measurement results. The longitudinal elongation distribution patterns are similar for all heads. The absolute value of the relative deviation (*ARD*) between the simulation and measurement, which is defined as,

Fig. 5.21 Measurement process: **a** Steps 1 and 2: drawing the markers; **b** Step 3: stamping the head; **c** Step 4: Measuring the positions of the markers on the formed head (from Ref. [17] with permission of Elsevier)

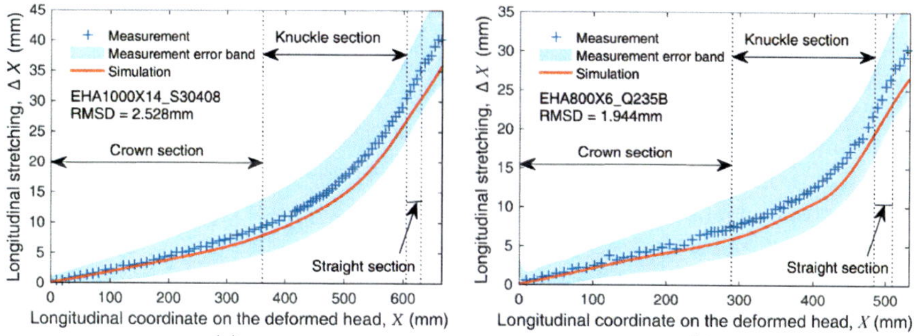

(a) (b)

Fig. 5.22 Comparison between simulated and measured data: longitudinal elongation distribution of longitudinal coordinates on deformed head. **a** EHA1000X14 made of S30408; **b** EHA800X6 made of Q235B (from Ref. [17] with permission of Elsevier)

$$ARD = \text{abs}\left(\frac{X_{\text{mean}} - X_{\text{sim}}}{X_{\text{mean}}}\right)$$

is presented as a longitudinal distribution in Fig. 5.23a, while the empirical cumulative distribution of the *ARD* is plotted in Fig. 5.23b. Generally, large *ARD* is observed in the crown section of the head because of the low absolute value of the

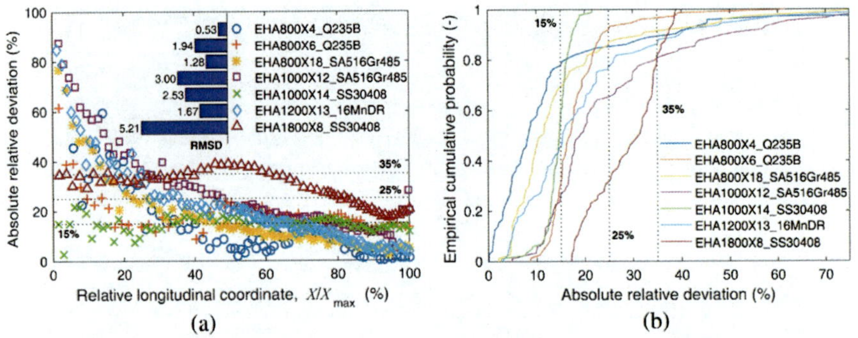

Fig. 5.23 Comparison between simulated and measured data: **a** Longitudinal elongation distribution in longitudinal coordinates on deformed head; **b** Empirical cumulative distribution of *ARD* (from Ref. [17] with permission of Elsevier)

data. Nonetheless, most of the measured data is located in the range of $ARD < 25\%$, providing good validation of the FE modeling. Thus, FE simulation can be used to analyze the detailed plastic deformation behavior of heads during stamping and predict the maximum PEEQ with respect to the head dimensions.

5.3.3 *Formula for Predicting Forming Strain*

By using the FE model described in Sect. 5.3.1, the detailed plastic deformation of heads during stamping was investigated. Extensive simulations with various head dimensions were carried out to find an empirical equation that could estimate the maximum plastic deformation during stamping based on head dimensions. The proposed empirical equation was then compared to the calculation equations suggested by the current codes and standards.

5.3.3.1 Detailed Plastic Deformation Behavior of Ellipsoidal Head

The FE simulation shows that complicated plastic deformation occurs during the stamping process of head forming. Taking EHA1000X14_S30408 as an example, Fig. 5.24 illustrates a sequence of the stamping process along with the PEEQ clouds on the material, while Fig. 5.25 presents the distribution of strains in three directions and PEEQ associated with the outer surface of the head.

At the very beginning ($Y_{\mathrm{punch}} = 0.057 D_{\mathrm{n}}$), the binder moves down to press the periphery region of the blank, bending the material outwards into an "S-step-shape (cross-sectional)" over the draw-bead. Correspondingly, wave-like shapes are generated on both the longitudinal strain distribution curve and thickness strain distribution curve, but with reversed directions, as indicated by "①" on the blue dotted lines in

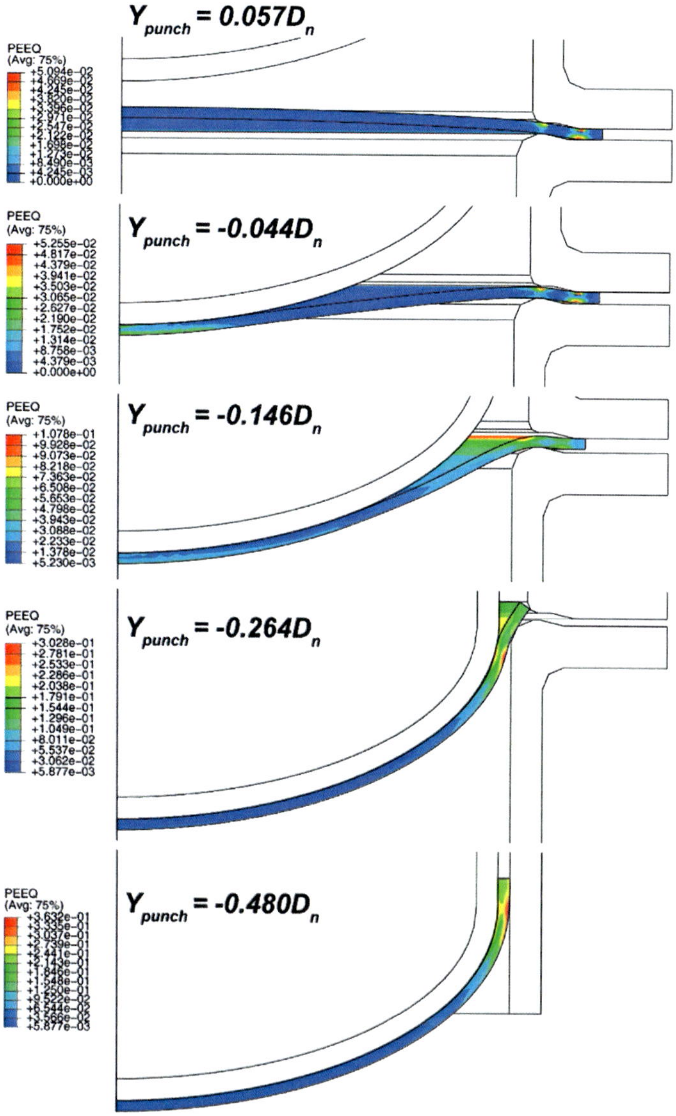

Fig. 5.24 PEEQ distribution on head (EHA1000X14_S30408) during stamping. Y_{punch} is convex die position which is zero when die touches blank (from Ref. [17] with permission of Elsevier)

Fig. 5.25a and c, indicating that the material is compressed in one direction while being stretched in the other, and there is no deformation in the third direction of latitudinal strain.

Next ($Y_{\text{punch}} = 0.057D_{\text{n}} - -0.044D_{\text{n}}$), the binder clamps the edge of the blank with a certain force and the punch moves down and touches the center of the blank, pushing the central material downwards and resulting in longitudinal/latitudinal

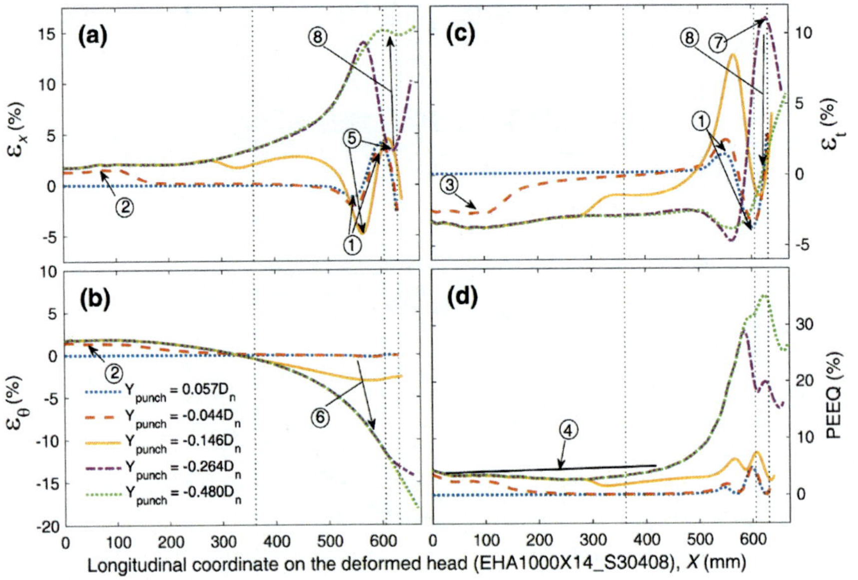

Fig. 5.25 Plastic deformation distribution on outer surface of formed head during stamping. Vertical axis: **a** Normal strain in longitudinal direction (ε_X); **b** Normal strain in latitudinal direction (ε_θ); **c** Normal strain in thickness direction (ε_t); **d** Equivalent plastic strain (PEEQ) (from Ref. [17] with permission of Elsevier)

stretching, labeled "②" on the red dashed lines in Fig. 5.25a and b. Positive strains in these two directions allow the material to flow out from the center and make the blank thinner, as indicated by "③" in Fig. 5.25c. Normal strains in three directions in the middle part of the blank (200 mm < X < 500 mm) remain at around zero at this moment, as shown in Fig. 5.25. This indicates that the material of this area is not yet deformed as it has not yet untouched the binder or punch.

As the punch goes down ($Y_{\text{punch}} = -0.044D_n - 0.264D_n$), the deformation clouds initiate and expand quickly from two areas (where $X = 0$ and $X \approx 550$ mm). Pushed out of the blank plane, the material in the crown section is generally stretched in both longitudinal and latitudinal directions, thinning it out. The deformation in the crown section stops once the material is fully wrapped on the punch, which is proven in Fig. 5.25 by the overlapping of the yellow, purple and green curves in the crown section (labeled "④" in Fig. 5.25d). The final strain level in the crown section is relatively low, and the PEEQ is less than 5%. On the other hand, the deformation of the outer part continually increases. As shown in Figs. 5.24 and 5.25a, the material in knuckle section and straight section is generally pulled in the longitudinal direction through the "S-shaped" gap between the binder and die. The outwards bending at the edge of the die compresses the outer surface material and deepens the valley on the $\varepsilon_X - X$ curve, labeled "⑤" in Fig. 5.25a. As the stamping proceeds, this "valley" moves outwards to the edge of the head. A similar movement of a "peak" will also be expected on the $\varepsilon_X - X$ curve for the inner surface material of the head. Moreover,

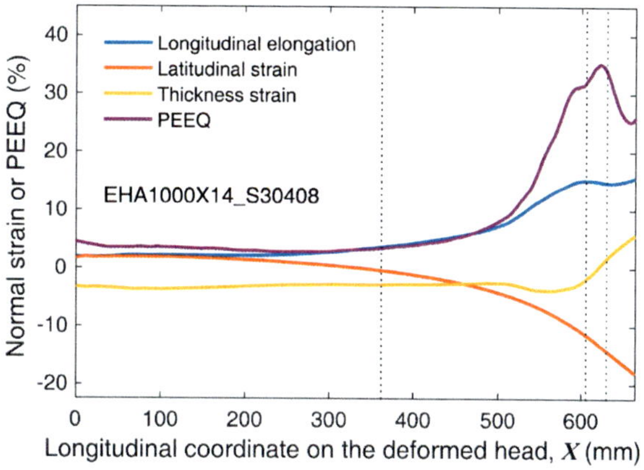

Fig. 5.26 Final deformation distribution on stamped head (EHA1000X14_S30408) (from Ref. [17] with permission of Elsevier)

the material in the knuckle section and straight section is wrapped inwards resulting in a significant compression in the latitudinal direction, labeled "⑥" on the $\varepsilon_\theta - X$ curve in Fig. 5.25b. The normal strain in the thickness direction is basically the result of the competition between the longitudinal stretching and latitudinal compression. The central part of the material is thinned out while the edge parts are thickened. A "peak" thickness strain can be found in the straight section at the end of this stage (labeled "⑦" in Fig. 5.25c).

In the final stage of stamping ($Y_{\text{punch}} = -0.264D_{\text{n}} - -0.480D_{\text{n}}$), the material is pulled out of the clamp by the binder and drops into the die. The thickened edge of the material then squeezes into the gap between the two dies where it is pulled and thinned. That is why the "valley" on the $\varepsilon_X - X$ curve and the "peak" thickness strain in the straight section disappear in this final stage (labeled "⑧" in Fig. 5.24c).

The final deformation of the head after stamping is presented in Fig. 5.26. Large PEEQ occurs at the periphery of the knuckle section and extends to the outer edge of the straight section, and a maximum of 35% is found in the middle of the straight section. At the edge of the material outside the straight section ($X > 0.606D_{\text{n}} + h_{\text{s}}$), the PEEQ increase again probably because of the free constraint at the edge. Nevertheless, this is of no interest as the material at the edge is the manufacturing margin which is cut off after the stamping process. A method for predicting plastic deformation distribution of cold formed standard ellipsoidal heads can be seen in references [18, 19].

5.3.3.2 Formula for Predicting Maximum Plastic Deformation

In practices, the maximum plastic deformation of a stamped head is used as a criterion to determine whether an additional heat treatment is required. For an m-ASS head, limiting the maximum deformation of the head is critical as excessive deformation induces large amounts of martensite formation, causing reductions in the toughness, ductility and corrosion resistance of the m-ASS. In this section, we present an empirical equation for calculating the maximum forming strain of a stamped ellipsoidal head based on extensive FE simulations.

By using FE simulation, extensive cases were simulated based on different head dimensions. Figure 5.27a and b illustrate the simulated maximum plastic deformation characterized by PEEQ during stamping as a function of the nominal diameter and the

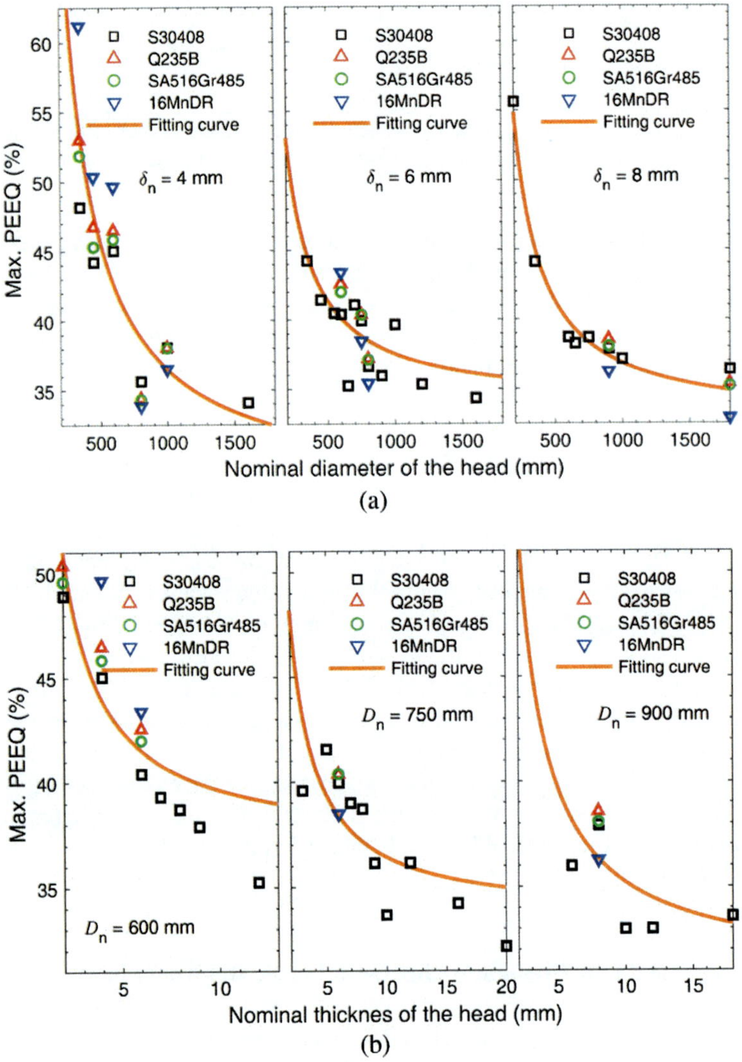

Fig. 5.27 Maximum plastic deformation (characterized by PEEQ) of stamped ellipsoidal head: **a** Versus nominal diameter with nominal thickness of 4/6/8 mm; **b** Versus nominal thickness with nominal diameter 600/750/900 mm (from Ref. [17] with permission of Elsevier)

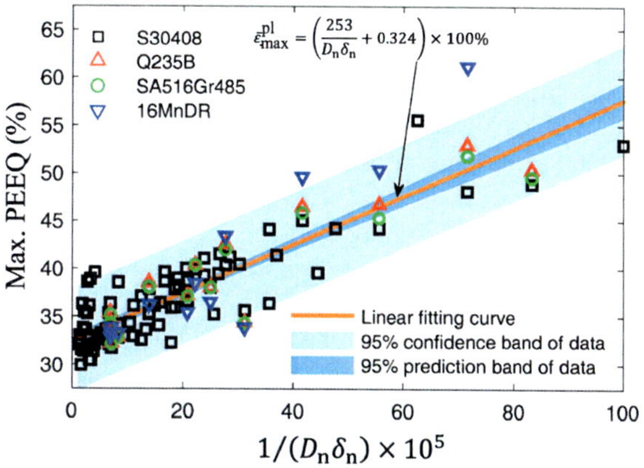

Fig. 5.28 PEEQ occurs on ellipsodial head during stamping forming as function of head dimensions (from Ref. [17] with permission of Elsevier)

nominal thickness respectively. The maximum PEEQ of a stamped ellipsoidal head is generally inversely proportional to both the nominal diameter and thickness of the head, whereas the material has a minor effect. Therefore, the maximum deformations were expected to be linearly related to $(D_n\delta_n)^{-1}$.

In Fig. 5.28, the maximum deformation for various heads is presented against $(D_n\delta_n)^{-1}$. A linear relationship is illustrated between the maximum deformation and $(D_n\delta_n)^{-1}$. An empirical correlation is generated by least square fitting, which gives,

$$PEEQ_{\max} = \bar{\varepsilon}^{\mathrm{pl}}_{\max} = \left(\frac{253}{D_n\delta_n} + 0.324\right) \times 100\% \quad (5.2)$$

Furthermore, a 95% confidence interval of the calculation results is also shown in Fig. 5.28, concluding an interval of $\bar{\varepsilon}^{\mathrm{pl}}_{\max} \pm 4.4\%$ for predicting the maximum deformation. Formula (5.2) is applicable to ellipsoidal heads made of more types of steel including stainless steel, carbon steel and low alloy steel compared with those developed by Zhu [7] and Ma et al. [20] which are only applicable for S30408.

5.3.4 Comparison of Different Formulas for Predicting Forming Strain

As described in Sect. 5.3.3.1, deformation in the knuckle section and straight section is complicated process: the inward wrapping bends the material in the longitudinal direction and generates significant compression in the latitudinal direction, while the pressing in the thickness direction constrains the thickening of the plate but forces the material to flow out in the longitudinal direction. All of these processes are highly coupled and none of them can be neglected.

The analytical Formula (1.52) presented by ASME VIII-1, GB/T 150.4 and GB/T 25198 calculates the maximum engineering strain of metal fiber in the longitudinal direction based on an analogy to a plate bending in one direction. According to Formula (1.52), the maximum deformation strain is considered to occur at the edge of the knuckle section (vertex of the ellipse on the major axis) where the curvature radius is maximum, i.e. $\rho_{\min} = 0.17D_n$. However, such a simplified assumption that the plate material is free of constrains on the surface is far beyond the realistic phenomenon in which the pressing action by the two dies on the material significantly constrains the thickening of the material and increases the longitudinal elongation. As expected, the calculated value by Formula (1.52) is much lower than the measurements and results from FE simulations, as shown in Fig. 5.29. Using Formula (1.52) recommended by ASME VIII-1, GB/T 150.4 and GB/T 25198 usually obtains underestimated forming strains, and essential heat treatment for many heads is not required and neglected as a result.

ASME VIII-2 and EN 13445–4 suggest using integrated true strain in the latitudinal direction (calculated by Formula (1.53)) at the edge of the head to characterize the maximum forming strain. The drawback of this method is obvious: the diameter of the blank, D_b, is unknown and arbitrary in real manufacturing. Using D_b in Formula (1.53) calculates the strain corresponding to the edge of the formed head. However, there are margins on the blank for stamping and the edge of the formed head will be cut. Thus, the maximum forming strain calculated on the basis of the cut part is no significance. In Fig. 5.29, the calculated values of the maximum forming strain

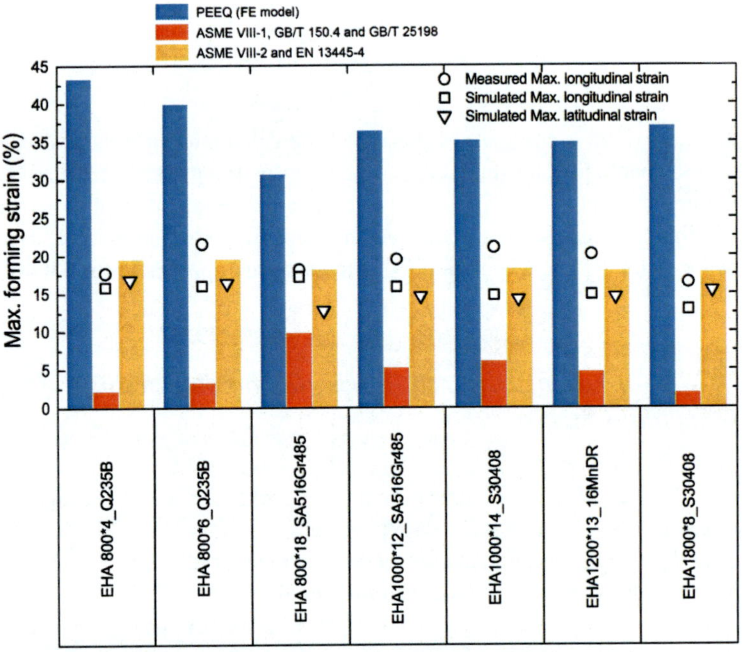

Fig. 5.29 Comparison between different methods for predicting the maximum forming strain of stamped ellipsoidal heads (from Ref. [17] with permission of Elsevier)

according to Formula (1.53) (yellow bars) are based on blank sizes with no margins that are given by the FE simulations. Furthermore, similar to Formula (1.52) recommended by ASME VIII-1, GB/T 150.4 and GB/T 25198, Formula (1.53) in ASME VIII-2 and EN 13445–4 only focuses on the strain in one direction and neglects highly coupled deformation in the other directions.

Moreover, PEEQ is a time-integrated value that records deformation behavior during the entire stamping process, while Formulas (1.52) and (1.53) only estimate the final deformation conditions. The reason for monitoring the deformation of stamping is to control the mechanical degradation of the material, which is more significantly related to the material's history of plastic deformation. As described in Sect. 5.3.3.1, the material in the knuckle and straight sections indeed experience compression first, then stretch in the longitudinal direction as shown in Fig. 5.25a. Both compression and stretching must be positively taken into account in order to reduce the mechanical performance of the material. Thus, PEEQ is the superior parameter for characterizing deformation during head stamping.

In this section, we emphasize that the plastic strain of a head during stamping forming is complex and consists of deformation in all directions. However, the equations recommended by the current codes and standards analytically describe the strain in only one direction, either longitudinal (ASME VIII-1, GB/T 150.4 and GB/T 25198) or latitudinal (ASME VIII-2 and EN 13445–4). Thus, PEEQ is proposed for characterizing complex plastic deformation during stamping forming of ellipsoidal heads. PEEQ obtained by FE analysis presents an accumulation of the plastic deformation during the stamping. It is therefore more realistic and accurate for interpreting how large plastic deformation degrades the mechanical properties of the material.

5.4 Forming Temperature

As mentioned in Sect. 5.1.3, forming temperature affects the cryogenic mechanical properties of material, especially enhancing its impact properties. Therefore, in this section, we investigated the appropriate forming temperature for the stamping of ellipsoidal heads. First, a series of tensile tests on austenitic stainless steel S30408 were conducted at elevated temperatures varying from 20 to 180 °C. Second, strain-induced martensitic transformation was investigated by tensile testing at elevated temperatures. Third, we developed a method for determining appropriate warm forming temperature to avoid strain-induced martensitic transformation. Finally, warm stamping for S30408 ellipsoidal heads was performed and compared with cold stamped heads. Section 5.4 is adapted from Refs. [16, 21] with permission of Elsevier.

5.4.1 Mechanical Properties at Different Temperatures

The material under investigation was obtained from four different batches of S30408 (corresponding to material grades 304 of ASTM A240/240 M-11a [22] and 1.4301 of EN 10028–7 [23]) m-ASS hot rolled plates with a thickness of 14 mm. All the plates were processed by solid solution treatment to eliminate strain-induced phases within the material. The four S30408 m-ASS plates are labeled I, II, III, and IV, and their chemical compositions are listed in Table 5.3 as compared with the requirements of the standards GB/T 24511 and EN 10028–7. It can be seen that chemical compositions of the four S30408 m-ASS plates meet the requirements of the standards GB/T 24511 and EN 10028–7.

For the tensile test, conventional cylindrical tensile specimens were machined from the I, II, III, and IV plates according to the Chinese standards GB/T 4338 [24] and GB/T 228.1 The nominal dimensions of the specimens in the gauge section were 25.0 mm (length) × 5.0 mm (diameter), as shown in Fig. 5.30.

Tensile tests at a constant strain rate of 3×10^{-4} s^{-1} and different temperatures varying from 20 to 180 °C were executed using an MTS810 standard testing machine equipped with a heating furnace. The strain was measured by a 632.53F-14 model high-temperature extensometer. Thermocouples were used for temperature measurements.

The engineering stress–strain curves obtained from the tensile tests at different temperatures are shown in Fig. 5.31. For variations in ductility, according to Fig. 5.31a and b, the elongation to fracture of S30408 m-ASS I and II was slightly saturated with the temperature increasing from 20 to 45 °C, probably due to their relatively high nitrogen fractions (see Table 5.3), while the elongation of S30408 m-ASS III and IV experienced a striking increase as the temperature increased from 20 to 45 °C. The elongation consistently reduced as the temperature increased from 45 to 180 °C. Highlighting the most obvious upsurge, the elongation value of S30408 m-ASS reached a peak of around 75% at 45 °C and bottomed out to around 50% at 180 °C. This indicates that the ductility properties of m-ASS weaken gradually as the temperature rose from 45 to 180 °C.

For variations in tensile strength, the measured ultimate strength (S_u) and yield strength ($R_{p0.2}$) are plotted against temperature from 20 to 180 °C in Fig. 5.32. As the temperature increased, the m-ASS strength degraded. It is clear that the values of S_u dropped more rapidly than those of the $R_{p0.2}$ as the temperature increased. With the higher strength and hardness of the strain-induced martensite (SIM) phase (mainly the BCC-α' phase) formed due to tensile tests than the parent phase of austenite (FCC-γ phase) in S30408, the SIM is usually considered as a source of the strengthening effect. Thus, concerning the strengthening effect from SIM and the only variable temperature set in these tests, it can be inferred that the temperature restraint effect on SIM is more significant as the temperature increases.

Table 5.3 Chemical composition of tensile test materials and standard values in Wt.% (from Ref. [21] with permission of Elsevier)

Material	C	Si	Mn	P	S	Cr	Ni	N	Mo
I	0.037	0.43	1.17	0.026	0.003	18.12	8.27	0.057	0.0563
II	0.038	0.42	1.08	0.025	0.003	18.42	8.02	0.052	0.1545
III	0.040	0.39	1.17	0.016	0.001	18.35	8.01	0.038	0.0058
IV	0.031	0.32	1.16	0.03	0.002	18.15	8.14	0.032	0.0711
GB/T 24511	≤ 0.08	≤ 0.75	≤ 2.00	≤ 0.035	≤ 0.020	18.00–20.00	8.00–10.50	≤ 0.10	-
EN 10028–7	≤ 0.07	≤ 1.00	≤ 2.00	≤ 0.045	≤ 0.015	17.50–19.50	8.00–10.50	≤ 0.10	-

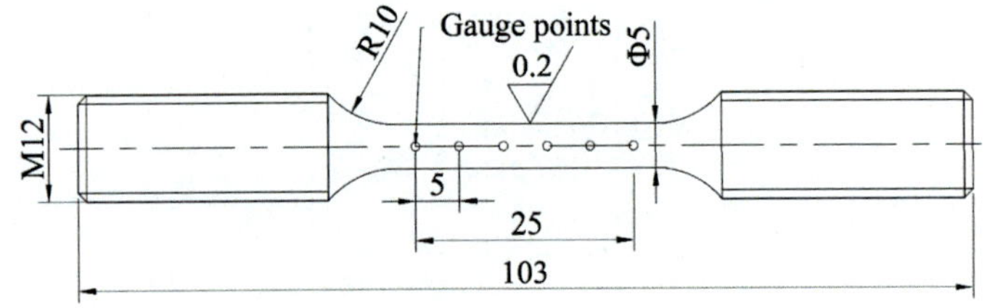

Fig. 5.30 Schematic illustration of tensile specimen with gauge points (Unit: mm) (from Ref. [21] with permission of Elsevier)

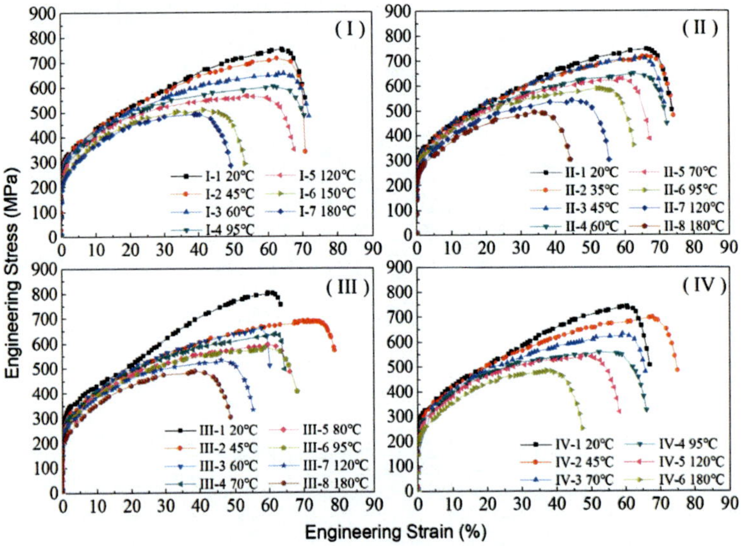

Fig. 5.31 Engineering stress–strain curves for materials **a** I, **b** II, **c** III, and **d** IV (from Ref. [21] with permission of Elsevier)

5.4.2 Strain-Induced Martensitic Transformation of Austenitic Stainless Steel

Due to the instability of its microstructure, m-ASS material can undergo strain-induced martensitic transformation (SIM-Tr) during the cold forming process. The SIM in m-ASS is basically BCC α'-phase which is strongly ferromagnetic and the only magnetic phase in the SIM-Tr of m-ASS [25]. Therefore, the ferrite number (*FN*) of specimens, measured by a Fisher MP30 ferritescope before and after the tensile tests, is applied to characterize the martensite amount. The ferritescope device was calibrated using delta-ferrite samples with a measurement range of 0.1–80% delta-ferrite prior to measurement. Each gauge point on the specimens (Fig. 5.30) was

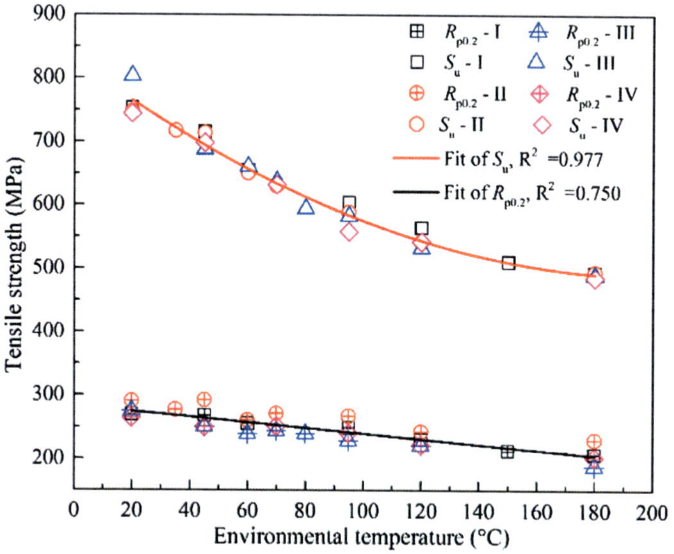

Fig. 5.32 Strength results of warm tensile tests at 20–180 °C (from Ref. [21] with permission of Elsevier)

measured with the ferritescope three times, then the average values of the ferritescope readings (*FN*) were taken as the test results of *FN* value.

For variations in martensite content, the *FN* values measured before and after the tensile test were plotted against temperature in Fig. 5.33. The initial *FN* before the tensile test was generally at a low level, within a range of 0.8–2.5. The *FN* increased after the tensile test, and the *FN* increased significantly by more than 20 after a room-temperature tensile test; however, the variation of *FN* was generally less than 4 at a tensile temperature of 95 °C and reduced to 0.4 when the temperature exceeded 120 °C. The increase in *FN* reduced with the increase in temperature because the increase in *FN* due to material deformation during the tensile test, i.e. SIM, was restrained by the rising temperature.

5.4.3 Method for Determining Warm Forming Temperature

As concluded in Sect. 5.4.2, warm forming temperature can significantly reduce the transformation (ΔFN) of the SIM during material deformation (Fig. 5.33). Therefore, we can develop a method for determining the appropriate warm forming temperature based on the ΔFN results from the tensile tests at temperatures from 20 to 180 °C.

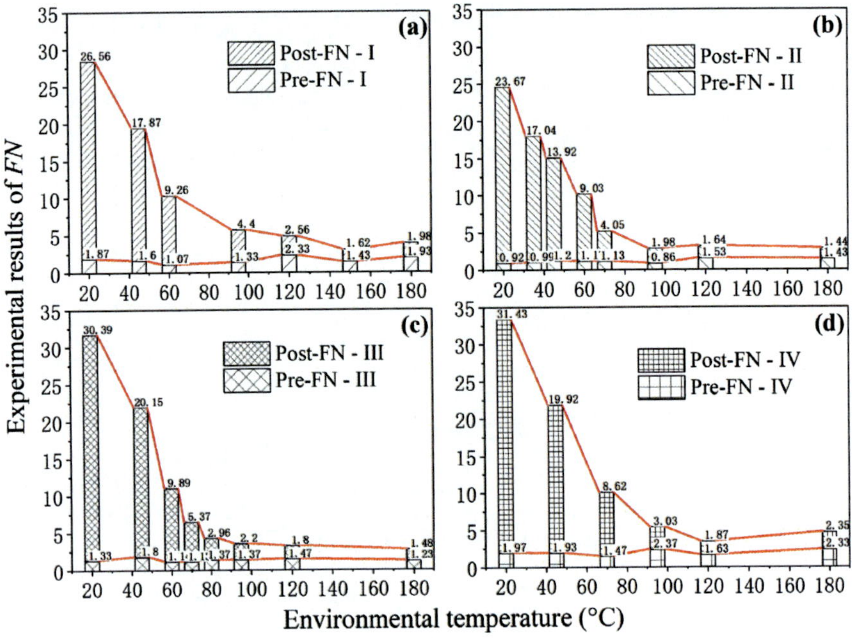

Fig. 5.33 Results of *FN* values for materials I, II, III and IV at 20–180 °C (from Ref. [21] with permission of Elsevier)

5.4.3.1 Calculation Method for ΔFN

Based on the *FN* measurement (Fig. 5.33) before and after the tensile tests, the SIM of the material during deformation, i.e. ΔFN, can be determined. Figure 5.34 plots ΔFN versus test temperature, and shows that ΔFN decreases exponentially as the temperature rises.

Therefore, the $\Delta FN - T$ curves were fitted on basis of the following correlation,

$$\Delta FN = FN_{\text{sat}}\exp(BT) \tag{5.3}$$

where FN_{sat} characterizes the saturation value of the amount of transformed martensite, B denotes the sensitivity of martensitic transformation to forming temperature, and T represents forming temperature in Celsius. As shown in Fig. 5.34, the ΔFN of materials after forming is basically eliminated when the forming temperature exceeds 120 °C. Focusing on the adjusted determination coefficient (R^2), it can be seen that the fitting result is reasonable due to the values of adjusted R^2 of the fitting curve fluctuates from 0.94 to 0.97, closer to 1. Thus, the fitting results are reasonable.

According to Formula (5.3), the ΔFN value of material during deformation can be predicted over a forming temperature period from 20 to 180 °C, with the value of

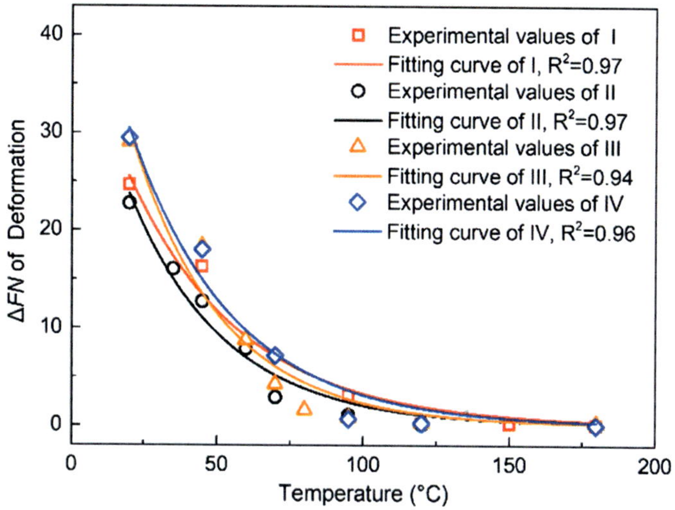

Fig. 5.34 Fitting curves of experimental results of ΔFN for materials I, II, III and IV (from Ref. [21] with permission of Elsevier)

FN_{sat} and B determined. The FN_{sat} and B are expected to be calculated by means of chemical composition, as the element types and contents are convenient to acquire.

Martensitic transformation occurs in two successive stages: nucleation for the formation of martensite and the growth of martensite [26]. Nucleation for the formation of martensite formation has been proved to be the key factor influencing the growth of martensite during SIM-Tr, and martensite growth depends on the increasing amount of nucleation and the mergers of martensite embryos [27–29]. Meanwhile, SIM-Tr is well known as a typical autocatalytic process [30], and the nucleation stage relies on special sites such as the intersections of slip bands [31]. In addition, stacking fault energy (*SFE*) is essential for characterizing defect structures such as dislocation cross slip, which significantly influence the mechanical and chemical properties. Therefore, *SFE* is the main factor in martensitic transformation as it is related to the formation of slip bands. The *SFE* value can then be calculated according to the following semi-empirical formula [32]:

$$SFE = -53 + 6.2\omega(\text{Ni}) + 0.7\omega(\text{Cr}) + 3.2\omega(\text{Mn}) + 9.3\omega(\text{Mo})\ \left(\text{mJ} \cdot \text{m}^2\right) \quad (5.4)$$

where $\omega(\text{Ni})$, $\omega(\text{Cr})$, $\omega(\text{Mn})$ and $\omega(\text{Mo})$ stand for the mass fractions of corresponding chemical compositions respectively. Thus, the *SFE* values of the materials I, II, III, and IV are obtained on basis of the corresponding chemical compositions in Table 5.3 according to Formula (5.4) (see Table 5.4).

To determine the two coefficients FN_{sat} and B, functions concerning the relation of chemical composition variation and coefficients need to be subsequently developed. Based on these results in Fig. 5.34, the two coefficients FN_{sat} and B are obtained

Table 5.4 *SFE* values and corresponding parameters FN_{sat} and *B* of materials (from Ref. [21] with permission of Elsevier)

Material	*SFE* ($mJ\cdot m^{-2}$)	FN_{sat}	*B*
I	15.23	42.87	−2.579E-02
II	14.51	43.75	−3.054E-02
III	13.30	57.87	−3.227E-02
IV	14.55	53.89	−2.851E-02
304	20.46	13.23	−1.981E-02
301	9.60	109.00	−4.130E-02

for each material I, II, III, and IV (see Table 5.4). As supplementary calculation examples, the values of FN_{sat}, *B*, and *SFE* of other m-ASSs such as 301 and 304 [10] were calculated according to Formulas (5.3) and (5.4), and the results are listed in Table 5.4. Thus, the difference in chemical composition explains the difference in *SFE* value for different m-ASSs. During the calculation, the volume fraction of martensite for m-ASSs 301 and 304 was transformed into the *FN* value by Formula (5.5) [33]:

$$FN = C_{\alpha'}/1.7 \tag{5.5}$$

where $C_{\alpha'}$ represents the actual mass fraction of the martensite phase. The *FN* value can be calculated approximately based on the mass fraction of the martensite phase, which is roughly the same as the volumetric fraction since the difference in density between martensite (7.761 $kg\cdot m^{-3}$) and austenite (7.911 $kg\cdot m^{-3}$) is only 2% [34].

The coefficients FN_{sat} and *B* were assumed to be calculated as functions of the *SFE* value, and the values of FN_{sat} and *B* were then plotted versus the *SFE* value respectively. The fitting results are shown in Fig. 5.35.

The fitting results of the relationships of $FN_{sat}-SFE$ and $B-SFE$ are reasonable because the values of the adjusted R^2 of fitting curves $FN_{sat} - SFE$ and $B - SFE$ are 0.98 and 0.93 (closer to 1) respectively. Thus, the coefficients of FN_{sat} and *B* are formulated as follows,

$$FN_{sat} = 308.20 - 26.46SFE + 0.589SFE^2 \tag{5.6}$$

$$B = -0.05866 + 0.00198SFE \tag{5.7}$$

Therefore, the ΔFN value is determined as a function of the chemical composition and forming temperatures of the materials by means of Formulas (5.3), (5.4), (5.6) and (5.7).

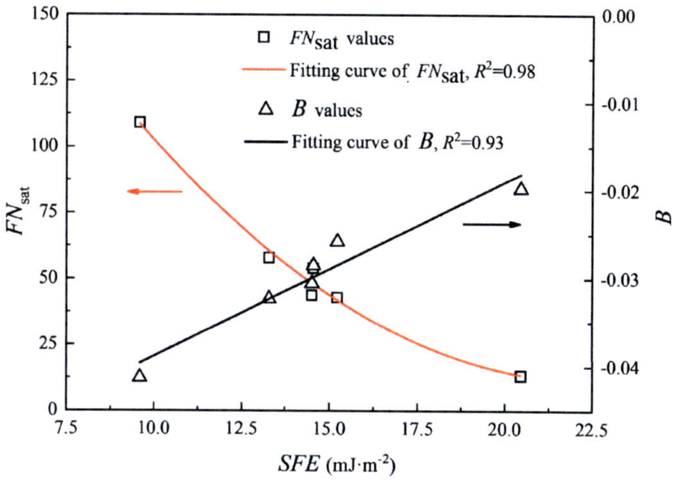

Fig. 5.35 Fitting curves of FN_{sat} and B values (from Ref. [21] with permission of Elsevier)

5.4.3.2 Proposed Forming Temperature

To determine the optimum forming temperature, the ΔFN values at different temperatures for a series of 304 m-ASS (Table 5.3) were calculated using Formulas (5.3), (5.4), (5.6) and (5.7), as shown in Fig. 5.36. The maximum and minimum values were fitted to provide the upper and lower bounds.

As shown in Fig. 5.36, ΔFN decreases significantly as the forming temperature rises from 70 °C to 180 °C. An ΔFN value below 3 is considered negligible since the pre-FN values of the tensile test specimens are generally less than 3.

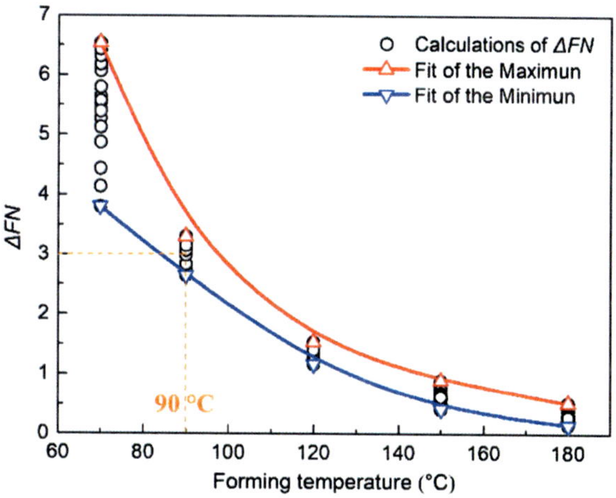

Fig. 5.36 ΔFN values of 304 m-ASS under various forming temperatures (from Ref. [21] with permission of Elsevier)

Table 5.5 Main chemical composition of S30408 m-ASS material in the stamping test (Wt. %) (from Ref. [21] with permission of Elsevier)

Head No	C	Si	Mn	P	S	Cr	Ni	N	Mo
CSH (EHA350 × 6)	0.044	0.35	1.10	0.024	0.006	18.50	8.03	0.048	0.0388
WFH (EHA1000 × 14)	0.048	0.53	1.12	0.028	0.004	18.09	8.06	0.076	0.0600

Thus, a forming temperature of 90 °C for m-ASS material is proposed by setting the maximum permitted ΔFN as 3.

5.4.4 *Advantages of Warm Forming*

Head stamping testing was carried out to verify the effectiveness of warm forming at the proposed forming temperature of 90 °C. The head material was slightly overheated before stamping to ensure that the measured head temperature after forming was above 90 °C but below 180 °C. The types of heads in this verification were EHA350 × 6 (WSH) presented in Sect. 5.1.3 and EHA1000 × 14 (WFH). The material for the head stamping test was S30408 treated m-ASS with solid solution, and the material chemical compositions for the two heads are listed in Table 5.5. EHA350 × 6 (WSH) and EHA1000 × 14 (WFH) were stamped at the temperatures of 90–120 °C and 90–158 °C respectively. In addition, the cold stamped head EHA350 × 6 (CSH) presented in Sect. 5.1.2 was stamped at room temperature (20–25 °C) for comparison.

5.4.4.1 Martensitic Transformation

The ferritescope was used to measure the transformed martensite content (ΔFN) for the stamping of the heads. For the convenience of comparison, the relative positions of measuring points marked on the surface of the circular plates used to stamp the heads were characterized by relative distance *d*, which is defined by the ratio of the distance from the center of the circular plate to its radius. $d = 0$ corresponds to the plate center (i.e. head apex), while $d = 1$ corresponds to the plate edge (i.e. head edge). The results of the measured ΔFN values were plotted versus relative distance *d*, as presented in Fig. 5.37.

The heads are intentionally divided into three parts for convenient description: crown (near HA), knuckle (near HA to HB) and straight section (HB to HC, height: 25 mm). Regarding the position of HB concerning head type, two HB (dash) identity lines are shown in Fig. 5.37: one for CSH and WSH (relative position 0.89), and the other for WFH (relative position 0.96). As shown in Fig. 5.37, the ΔFN in the crown ($0 \leq d \leq 0.6$) is very low and basically remains constant with no apparent relationship to forming type (cold or warm stamping). As the plastic strain is almost

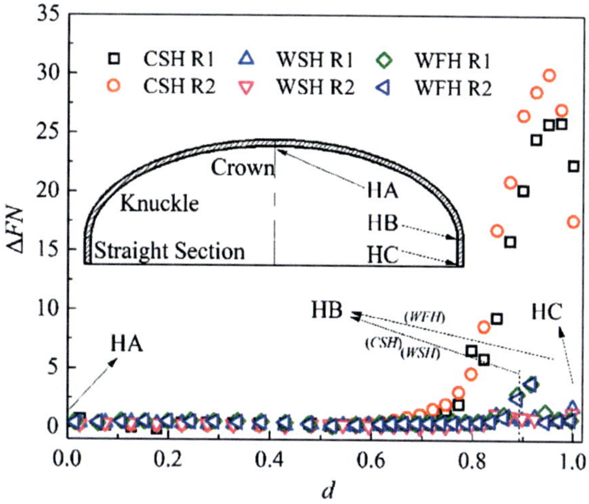

Fig. 5.37 ΔFN values of cold stamped head EHA350 × 6 (CSH) and warm stamped heads EHA350 × 6 (WSH) and EHA1000 × 14 (WFH) (from Ref. [21] with permission of Elsevier)

negligible in the crown, the inconspicuous SIM-Tr leads to the low SIM content. The ΔFN of CSH rises dramatically in the knuckle section of the head, corresponding to a relative position of 0.7–0.9, whereas the ΔFN of WSH and WFH remain almost constant in the knuckle section. For CSH, the ΔFN rises significantly as d increases close to the straight section of the head, i.e. $0.9 < d < 1.0$, and the ΔFN reaches a peak of more than 27. For WSH and WFH, the ΔFN rises slightly from near HB to HC as compared with CSH, and increases to around 1.78 and 3.93 in the straight section respectively. This indicates that compared with the m-ASS head cold stamped at room temperature, the SIM-Tr is significantly suppressed by warm forming with an appropriate temperature of over 90 °C.

The ΔFN values of CSH, WSH, and WFH at corresponding temperatures were calculated using the ΔFN calculating Formulas (5.3), (5.4), (5.6) and (5.7) proposed in Sect. 5.4.3. The calculations were then compared with the head stamping tests results, as listed in Table 5.6. For CSH and WSH, the maximum measured values of ΔFN are found exactly between the calculations of ΔFN. Although the maximum measured value of ΔFN for WFH exceeded the predicted range, the difference between them is small and the percent error of this prediction is approximately 16.8%, which can be accepted in engineering applications. Therefore, the consistency of ΔFN between measurements and predictions exemplifies the validity of

Table 5.6 Results of ΔFN values of head stamping test and calculation (from Ref. [21] with permission of Elsevier)

Forming type	Cold stamped head EHA350 × 6 (CSH)	Warm stamped head EHA350 × 6 (WSH)	Warm stamped head EHA1000 × 14 (WFH)
T (°C)	20.0–25.0	90.8–120.4	90.2–158.5
ΔFN Measured	30.05 (max)	1.78 (max)	3.93 (max)
ΔFN Predicted	25.86–30.30	1.25–3.21	0.38–3.27

the aforementioned formulas well, and the ΔFN of the S30408 m-ASS head can be controlled by improving the forming temperature as compared with that of CSH. Although the significant difference in *FN* between WSH and CSH can be explained by the SIM-Tr being suppressed at elevated temperatures, its effects on mechanical performance still require further study.

Furthermore, to investigate the influence of warm forming on the microstructures of m-ASS, the microstructural differences between WSH and CSH were examined by means of optical metallography. Cross-sections of the test samples were cut from the straight section of the heads using an electrical discharge machining (EDM). The samples were first mechanically polished using SiC emery paper up to a roughness of 2000 grit, then 1.5 μm diamond paste to a final roughness of ~0.1 μm. To reveal the grains, the sample surface was etched with a mixed solution of 5 g of $FeCl_3$, 25 ml of HCl, and 25 ml of ethanol for 5 s under ambient temperature. After preparation, the specimens were analyzed with a Keyence VHX-1000 microscope. The optical microscope images from the straight sections of CSH and WSH are shown in Fig. 5.38.

For the microstructure of CSH, as shown in Fig. 5.38a, it is confirmed that the grain structure consists of primarily austenitic grains (light colored areas), quite a few martensites (dark colored areas) and some elongated delta-ferrites (black strips). In the microstructure of WSH, as shown in Fig. 5.38b, the martensite phase is uniformly dispersed and formed on shear bands of the austenite phase. However, the amount of delta-ferrite in the WSH material is less than that in the CSH material, which explains why the measured ΔFN value of WSH is much less than that of CSH, as shown in Fig. 5.37 and Table 5.6). Therefore, it can be inferred that forming temperature can affect the precipitation of delta-ferrite during stamping.

It should be noted that, because many other areas may have consisted of an excessive etching solution that was distributed non-uniformly and might not have indicated the real martensite phase, no martensite content analysis was conducted. Notwithstanding its limitation, the microstructure comparison between WSH and CSH is significant for revealing that forming temperature may affect the quantity of delta-ferrite, whereas pitting corrosion easily occurs at the crossing points of the

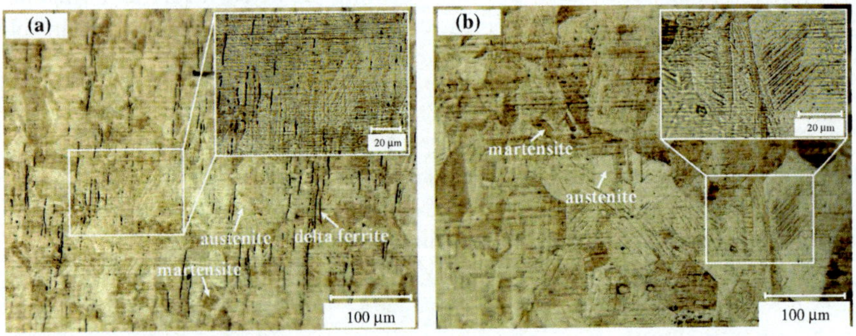

Fig. 5.38 Microstructures of straight section of **a** CSH and **b** WSH (from Ref. [21] with permission of Elsevier)

grain boundaries of delta-ferrite and austenite in m-ASS due to chromium depletion [34]. Meanwhile, this study indicates that a forming temperature of 90 °C is recommended for m-ASS as compared with cold forming. Therefore, it is of great interest to make further investigations to explore the effects of warm forming on the corrosion resistance properties of m-ASS in the future.

5.4.4.2 Cryogenic Impact Properties

The good cryogenic impact properties of m-ASS are the primary reason for its application. As concluded in Sect. 5.1.3, the cryogenic impact properties of WSH are greater than those of CSH, especially the material in the straight section (Fig. 5.7); thus, warm stamping can greatly enhance the cryogenic impact properties of head material.

In summary, the warm forming of m-ASS heads has two advantages: (1) A lower quantity of delta-ferrite in WSH was observed than that in CSH. (2) Warm stamping can enhance the cryogenic impact properties, as compared with cold stamping at room temperature. For m-ASS head stamping, a warm forming temperature of 90 °C is recommended. This recommended forming temperature can also be applied in the manufacturing of other m-ASS components such as elbows.

References

1. Wang H, Yao X, Li L, Sang Z, Krakauer BW (2017) Imperfection sensitivity of externally-pressurized, thin-walled, torispherical-head buckling. Thin-Walled Struct 113:104–110
2. Wan L, Tao W, Wu X, He S (2004) The influence of initial geometric imperfection on the localized buckling of pressure vessel under internal pressure. In: Proceedings of 12th international conference on nuclear engineering, Paper No. ICONE12–49397, pp 329–333.
3. Blachut J, Galletly GD (1993) Influence of local imperfections on the collapse strength of domed end closures. Proc Inst Mech Eng 207:197–207
4. Li Y, Bu F, Kan W, Pan H (2013) Deformation-induced martensitic transformation behavior in cold-rolled AISI304 stainless steels. Mater Manuf Process 28:256–259
5. Xu CC, Zhang XS, Hu G (2003) Effect of plastic deformation on texture and corrosion resistance of AISI304 stainless steel. CIESC J 54:790–795
6. Zhu X, Miao C, Ma L (2013) A method for prediction of forming strain for cold stamping formed head made of S30408 austenitic stainless steel. In: ASME PVP conference, PVP2013–97768
7. Zhu X (2014) Research on prediction and test method of plastic deformation for cold stamping formed standard elliptical head made of austenitic stainless steel. Master's Thesis, Zhejiang University, Hangzhou. (in Chinese)
8. Pang JE, Guo XH (2012) Analysis of cracking on stainless steel head in low temperature and hydrogen service. Process Equip Piping 49:64–67 (in Chinese)
9. Bong HJ, Barlat F, Ahn DC, Kim H, Lee M (2013) Formability of austenitic and ferritic stainless steels at warm forming temperature Int. J Mech Sci 75:94–109
10. Peterson SF, Mataya MC, Matlock DK (1997) The formability of austenitic stainless steels. JOM-Us 49:54–58
11. Huang GL, Matlock D, Krauss G (1989) Martensite formation, strain rate sensitivity, and deformation behavior of type 304 stainless steel sheet. Metall Trans A 20:1239–1246

12. Zheng J, Li K, Liu S, Ge H, Zhang Z, Gu C, Qian H, Hua Z (2018) Effect of shape imperfection on the buckling of large-scale thin-walled ellipsoidal head in steel nuclear containment. Thin-Walled Struct 124:514–522
13. Li K (2019) Research on buckling behavior and prediction method of large-scale thin-walled ellipsoidal head under internal pressure. PhD Thesis, Zhejiang University, Hangzhou. (in Chinese)
14. General Administration of Quality Supervision, Inspection, and Quarantine of the People's Republic of China, Standardization Administration of the People's Republic of China (2007) GB/T 13239–2006 Metallic materials—tensile testing at low temperature. Standards Press of China, Beijing
15. General Administration of Quality Supervision, Inspection, and Quarantine of the People's Republic of China, Standardization Administration of the People's Republic of China (2007) GB/T 229–2007 Metallic materials—charpy pendulum impact test method. Standards Press of China, Beijing
16. Zhang X (2015) Research on temperature of warm stamping formed standard elliptical head made of metastable austenitic stainless steel. Master's Thesis, Zhejiang University, Hangzhou. (in Chinese)
17. Zheng J, Shu X, Wu Y, Xu H, Lu Q, Liao B, Zhang B (2018) Investigation on the plastic deformation during the stamping of ellipsoidal heads for pressure vessels. Thin-Walled Struct 127:135–144
18. Wang K (2015) Research on residual effects of cold stamping on elliptical head and its characterization method. Master's Thesis, Zhejiang University, Hangzhou. (in Chinese)
19. Hui P, Wang K, Zheng J, Gu X, Ye Y (2016) Prediction method of plastic deformation distribution of cold stamping formed S30408 standard elliptical head. Light Ind Mach 34(5):27–33 (in Chinese)
20. Ma L, Miao C, Zhu X, Zheng J, Gu X, Ye Y (2015) Research on prediction method of plastic deformation for cold stamping formed standard elliptical head made of austenitic stainless steel. J Mech Eng 6:19–26 (in Chinese)
21. Zheng J, Lu Q, Wu Y, Zhang X, Ding H, Hui P, Li Q (2019) Research on forming temperature of metastable austenitic stainless steel head based on strain-induced martensitic transformation. ASME J Press Vessel Technol 141:051401
22. Committee ASTM (2014) ASTM A240/A240M: Standard specification for chromium and chromium-nickel stainless steel plate, sheet, and strip for pressure vessels and for general application. ASTM, West Conshohocken
23. CEN National Members (2008) EN 10028–7: Flat products made of steels for pressure purposes—part 7: stainless steels. Beuth Press Ltd., Berlin.
24. General Administration of Quality Supervision, Inspection, and Quarantine of the People's Republic of China, Standardization Administration of the People's Republic of China (2006) GB/T 4338–2006: Metallic materials tensile testing at elevated temperature. Standards Press of China, Beijing.
25. Meszaros I, Prohaszka J (2005) Magnetic investigation of the effect of α′-martensite on the properties of austenitic stainless steel. J. Mater Process Technol 161:162–168
26. Shin HC, Ha TK, Chang YW (2001) Kinetics of deformation induced martensitic transformation in a 304 stainless steel. Scr Mater 45:823–829
27. Huang J, Ye X, Xu Z (2012) Effect of cold rolling on microstructure and mechanical properties of AISI 301LN metastable austenitic stainless steels. J Iron Steel Res Int 19:59–63
28. Shrinivas V, Varma SK, Murr LE (1995) Deformation-induced martensitic characteristics in 304 and 316 stainless steels during room-temperature rolling. Metall Mater Trans A 26:661–671
29. Murr LE, Staudhammer KP, Hecker SS (1982) Effects of strain state and strain rate on deformation-induced transformation in 304 stainless steel: part II. Microstructural study. Metall Trans A 13: 627–635
30. Guntner CJ, Reed RP (1962) The effect of experimental variables including the martensitic transformation on the low temperature mechanical properties of austenitic stainless steels. Trans ASM 55:399–419
31. Olson GB, Cohen M (1975) Kinetics of strain-induced martensitic nucleation. Metall Trans A 6:791–795

32. Schramm RE, Reed RP (1975) Stacking fault energies of seven commercial austenitic stainless steels. Metall Trans A 6:1345–1351
33. Talonen J, Aspegren P, Hänninen H (2004) Comparison of different methods for measuring strain induced α-martensite content in austenitic steels. Mater Sci Techol Lond 20:1506–1512
34. Garfias-Garcia E, Colin-Paniagua FA, Herrera-HernáNdez H, Juarez-Garcia JM, Palomar-Pardave ME, Romero-Romo MR (2010) Electrochemical and microscopy study of localized corrosion on a sensitized stainless steel AISI 304. ECS Trans 29:93–102
35. Zhang Z (2021) Research on failure mechanism-based method for design of steel ellipsoidal heads under internal pressure. PhD Thesis, Zhejiang University, Hangzhou.

Chapter 6
Summary

This book is intended to provide comprehensive coverage of the stress, failure, design and fabrication of ellipsoidal heads for pressure vessels, which are summarized as follows:

Chapter 1 Introduction

Chapter 1 first addresses the various types of heads including flat, conical, shallow spherical, hemispherical, torispherical and ellipsoidal. The ellipsoidal head is the ideal head choice for pressure vessels due to its good stress distribution and ease of fabrication. The most widely used steels and their properties are then described. Ellipsoidal heads have wide applications in the chemical, petroleum, nuclear, marine, aerospace and food processing industries.

Stress analysis of ellipsoidal heads is performed using elastic or elastic–plastic theory, and compared with those of torispherical heads. According to the stress analysis, bi-axial tensile stresses in the head crown cause the occurrence of plastic collapse, while compressive circumferential stresses in the head knuckle may cause the occurrence of buckling. Thus, two types of failure mode are considered: buckling and plastic collapse. We then review the progress of research on the failure modes, as well as formulas for predicting the buckling pressure and plastic collapse pressure of ellipsoidal heads respectively.

This book provides a detailed review of the design methods of ellipsoidal heads in the pressure vessel codes and standards including ASME VIII-1 and VIII-2, GB/T 150.3, JB 4732 and EN 13445–3. It also reviews fabrication methods including pressing, spinning and assembly from formed segments, as well as fabrication requirements including shape deviation and forming strain. Finally, we review the design methods of special ellipsoidal heads including heads under external pressure, heads with variable thicknesses, and heads with nozzles.

Chapter 2 Buckling of Ellipsoidal Heads

In Chap. 2, we conduct an in-depth investigation into the buckling of ellipsoidal heads using experiments and nonlinear finite element method. In order to perform

J. Zheng and K. Li, *New Theory and Design of Ellipsoidal Heads for Pressure Vessels*,
https://doi.org/10.1007/978-981-16-0467-6_6

buckling experiments of the large test heads with a diameter up to 5 m, we designed reusable test vessels which permit the tested heads to be pressurized until rupture. Modern measurement technologies such as 3D laser scanning are used to measure the initial shape and deformation of ellipsoidal heads. Large strains under hydraulic pressure are measured using designed seal methods: strain gauges on the inside surfaces are waterproofed with silicone rubber, and sealed feedthrough assemblies are specially designed. Models of ellipsoidal heads are generated using the nonlinear finite element method, and perfect and actual shapes are both considered.

It is found that the buckling of ellipsoidal heads under internal pressure has three characteristics: locality, progressivity and self-limitation. Buckles first form at local positions on the head knuckle. The reason for the occurrence of buckling is that the head knuckle is under compressive stresses as it deforms inward. As the pressure increases, more buckles form around the circumference of the knuckle, which is a progressive process. Moreover, the growth of buckles is progressive but not sudden for some heads. The buckles become smaller and even disappear at relatively higher pressures because the compressive stresses in the head knuckle become smaller or even change to tensile stresses. Furthermore, the effects of material properties, radius-to-height ratio, diameter-to-thickness ratio and shape imperfection on the buckling of ellipsoidal heads are investigated. It is found that plastic buckling occurs in the investigated ellipsoidal heads, and yield strength increases buckling pressure. Material strain hardening can increase the resistance to buckling of ellipsoidal heads with smaller diameter-to-thickness ratios, but has no or little effect on the buckling pressure of ellipsoidal heads with larger diameter-to-thickness ratios. As the radius-to-height or diameter-to-thickness ratios decrease, buckling pressure increases, and the growth of a buckle changes from sudden to gradual, even no buckling occurs in ellipsoidal heads. Shape imperfection also promotes the formation of buckling.

In order to determine whether ellipsoidal heads will buckle or not, a new buckling criterion is developed. A new formula for predicting the buckling pressure is also proposed for ellipsoidal heads at which buckling occurs. The predictions of the new formula are in reasonably good agreement with the experimental results. Compared with other formulas, the new formula has comprehensive advantages in both accuracy and applicability.

Chapter 3 Plastic Collapse of Ellipsoidal Heads

In Chap. 3, we conduct a detailed analysis of the plastic collapse of ellipsoidal heads under internal pressure. Plastic collapse experiments are performed on the heads covering diameter-to-thickness ratios of 88–980, radius-to-height ratios of 1.7–2.4, many kinds of steel and three kinds of usual manufacturing methods (cold pressing, cold spinning, and assembly from formed segments). Reusable test vessels for the plastic collapse experiments of ellipsoidal heads were designed using the same method as those for the buckling experiments. We develop finite element models for calculating the plastic collapse pressure of ellipsoidal heads, taking into account the effects of material strain hardening and geometric nonlinearity.

The results show that ellipsoidal heads tend to become hemispherical, which indicates that ellipsoidal heads experience geometric strengthening before the occurrence of plastic collapse. Plastic collapse is the maximum load that an ellipsoidal head can carry, and ductile rupture occurs after plastic collapse. Furthermore, we investigate the effects of geometrical and material parameters on plastic collapse pressure. It is clear that engineering ultimate tensile strength and thickness-to-diameter ratio significantly increase plastic collapse pressure.

We develop a new simple formula for predicting the plastic collapse pressure of ellipsoidal heads. This formula is applicable to ellipsoidal heads with a radius-to-height ratio of 1.5–2.5 and diameter-to-thickness ratio of 20–2000, and which are made of steels specified in pressure vessel codes and standards, including stainless steel, carbon steel and low alloy steel. Compared with other predictions based on elastic or perfectly-plastic theory, the new simple formula provides better prediction of the plastic collapse pressure of ellipsoidal heads by virtue of taking material strain hardening into consideration.

Chapter 4 New Method for Design of Ellipsoidal Heads

In Chap. 4, the main problems affecting the methods for the design of ellipsoidal heads in the current codes and standards are discussed: (a) the current local buckling criteria are more conservative, limited and/or inaccurate; (b) the current design rules for buckling failure based on geometric equivalency approaches are on the conservative side, resulting in uneconomical design; and (c) the effect of material strengthening is not considered in the current design rules for plastic collapse, also leading to conservative design.

Therefore, we develop a new failure mechanism-based design method for ellipsoidal heads under internal pressure. Based upon the formulas for predicting buckling pressure and plastic collapse pressure, we respectively develop design formulas for calculating the minimum required thickness to prevent the buckling and plastic collapse of ellipsoidal heads. In addition, the new local buckling criterion is used to determine whether buckling is considered or not. If buckling is not considered, the required thickness of ellipsoidal heads is finally determined as the minimum required thickness to prevent plastic collapse. If buckling is considered, the minimum required thickness of ellipsoidal heads is determined as the maximum value between the minimum required thicknesses to prevent plastic collapse and buckling.

Compared with the current design methods in the codes and standards, the new method for the design of ellipsoidal heads has the following advantages: (a) the applicability scope of the new design method is broader than those of ASME VIII-2 and EN 13445–3; (b) the new local buckling criterion is more accurate and reasonable than the others, making it more applicable to determine whether or not internally pressurized ellipsoidal heads will buckle; and (c) the new method can provide smaller required thicknesses than the current codes and standards including ASME VIII-1, ASME

VIII-2, GB/T 150.3, JB 4732 and EN 13445–3, which gives it an economical advantage because the effects of material strain hardening and geometric strengthening are considered.

Chapter 5 Control of Fabrication Quality of Ellipsoidal Heads

In Chap. 5, we study controlling on the fabrication quality of ellipsoidal heads, including shape deviation, forming strain and forming temperature.

We investigate the effects of shape deviation on the buckling pressure of ellipsoidal heads under internal/external pressure. It is clear that shape deviation significantly decreases the buckling pressure of ellipsoidal heads, so it should be controlled in order to prevent the buckling failure of actual ellipsoidal heads. Therefore, we develop a method for evaluating the shape deviation of ellipsoidal heads on the basis of non-contact measurement, with which shape deviations can easily be obtained and characterized.

Pre-strain tensile and head stamping tests are performed to investigate the effects of forming strain on the mechanical properties of material. It is clear that forming strain affects the mechanical properties of material, especially reducing its impact properties. According to the pressure vessel codes and standards, forming strain is an essential criterion for deciding whether the post-forming heat treatment is required. Therefore, an accurate prediction of the plastic deformation of heads during stamping is a vital aspect of controlling head quality. In this book, the stamping process is simulated using the finite element method, and experimentally verified by comparing the simulation to the real longitudinal elongation on the stamped heads. The detailed plastic deformation behavior of ellipsoidal heads during stamping is presented. An empirical formula for predicting the maximum equivalent plastic strain (PEEQ) is developed. A comparison of the empirical formula with the formulas provided by the current codes and standards is discussed, showing that PEEQ is the superior parameter for characterizing deformation during head stamping.

Cold and warm stamping tests of ellipsoidal heads are performed. It is clear that forming temperature affects the cryogenic mechanical properties of material, especially enhancing its impact properties; thus, we investigate the appropriate forming temperature for the stamping of ellipsoidal heads. A series of tensile tests on metastable austenitic stainless steel (m-ASS) are conducted at elevated temperatures varying from 20 to 180 °C, and the corresponding strain-induced martensitic transformation is investigated. A method is developed for determining the appropriate warm forming temperature so as to avoid the strain-induced martensitic transformation. The warm forming of m-ASS heads has two advantages: (a) a lower quantity of delta-ferrite in warm stamped heads than in cold stamped heads is observed; and (b) warm stamping can enhance the cryogenic impact properties compared with cold stamping. For m-ASS head stamping, a warm forming temperature of 90 °C is recommended. This recommended forming temperature can be also applied to the fabrication of other m-ASS components such as elbows.